HOW TO BUDGET &

MANAGE YOUR

MONEY

Financial Planning Books for Beginners.
How to Save Money Faster, Pay Off Debt
And Control Your Finances

by

Rachel Mercer

TABLE OF CONTENTS

DISCLAIMER

Please note that this book should be used as educational tools only and are not replacement for professional investment advice. Neither the author, nor publisher is providing investment, financial, legal or tax advice, and nothing in this book should be construed as such by you. It is intended to provide only general information on personal finance, building wealth, paying down debt, and other topics. If you need professional advice, you should consult your own professional advisers and appropriate licensed professional. This book does not provide complete information on the subject matter covered. It is intended to be used only as a general guide, and not as a sole source of information on the subject matter.

While the author has undertaken diligent efforts to ensure accuracy, there is no guarantee of accuracy or of no errors, omissions or typographical errors. The views expressed are those of the author alone, and should not be taken as expert instruction or commands. The reader is responsible for his or her own actions.

YOUR FREE GIFT

Thanks for buying my book!

As a way of showing my appreciation, I'd like to give a **FREE bonus gift** exclusive to my readers.

I've created a special PDF report, **"11 Best Side Hustles You Can Do Anywhere at Any time in 2020 to Make Extra Money".** It will give you a simplest and easiest side hustle ideas to increase your income and tips on how to grow your wealth. Earning extra money these days has become easier than ever. You'll be surprised as to what you'll find out in this guide – as side hustles can definitely level up your finances.

These little things every month will go a long way. In several months or in a few years' time, you'll be surprised as to how much you had actually saved through little increments that accumulated into something bigger.

Get it Here:
https://bit.ly/39hoopf

You can also download and print worksheets that are shown in several chapters and use them. Please access free downloads by going here: https://BookHip.com/GPFXRX

INTRODUCTION

All of us race behind money to attain "financial success." But what exactly is financial success? Is it about having more assets to your name? Is it about having no debts? Or is it about having more savings? Is it a lucrative investment portfolio? The perception of financial success varies from person to person.

But what is the one common factor that drives you towards this financial success? Allow me to introduce you to the concept of budgeting. It does not matter if your definition of financial success differs from that of your friend's, for both of you can use the tool of budgeting to achieve this success. In simple words, budgeting is the process of being aware of your income and expenses. This knowledge helps you to make conscious financial decisions.

This book will show you how to **save money and manage finances with a personal budget plan** that works for you.

This book will show you how to **budget easily** and effectively so you can **stick to it** and practice effortlessly.

This book will give you **practical and realistic strategies** and ideas that will actually work for you and help you solve your financial problems.

This book will help you to be relaxed, secured, **relieved of stress**, and confident about your finances.

Taking control of your finance and being frugal is the most important and crucial component of becoming financially independent. By being frugal, you can control both sides of your finances—income and expense. You can reduce your expenses and save more at the same time while investing in assets that can provide passive income.

Budgeting is the first step to realizing your dream of financial success. Even the richest men budget; actually, they budget forever and are experts at it. They value and respect money. Warren Buffet, who is one of the world's richest men, known to have a net worth of over $85 billion, amassed that wealth by keeping his spending significantly lower than his expenses and investing the difference. With compound interest and great returns, his investments created the massive amount of wealth he has today.

The first step is to budget and be frugal. Frugality might sound negative and hard to follow, and sticking to a budget is not easy. But if you learn how to do it, and make a habit of it, it is actually not that difficult. You will be more relaxed and secured and feel confident. Needless to say, you will be with much less stress. Budgeting is taking control of your money, not the other way around, and taking responsibility for your finances.

Budgeting and being frugal also requires self-discipline. You need to learn to control your emotions. But this healthy habit will reward you far more than you expect and lead to an abundant and fulfilled life in the future.

This book is for you if:

- You are having a hard time paying your bills and saving your money.
- You want to start saving and budgeting but don't know where to start.

- You want a simple and easy to use budgeting plan that you can stick to.
- You want to walk your talk and be a person who takes action.
- You want to become a responsible person who accomplishes financial goals.
- You want to control your money, your life, and your financial destiny.

Establishing a budget is the act of deciding how much of your money you're going to spend on one item, how much on another, and so on, before you're actually in the position of spending the money. Sticking to a budget is the act of following through on those decisions. Creating a budget is easy, but *sticking to any budget* is very difficult.

The trick is to focus on the word **Realistic**. It doesn't take much research or many difficult decisions to decide that you're going to spend $200 per month on food. But if you've never spent less than $500 per month on food, you'll blow your budget right out of the water the first week. Instead, before you begin deciding on the numbers in your budget, you'll need to fully assess your current situation, take a hard look at where you can cut back your financial obligations (both large and small), restructure your debt (if necessary), and see whether you can add income. Only then are you ready to decide realistically where every penny will be spent.

A budget is a tool, and like all tools, the results you get from it will be determined by how you use it. If you make a realistic budget and stick to it, you can watch your life move forward. If you set unrealistic budgetary expectations and don't even bother to follow through with them, don't think your financial problems are over.

> *"Knowing is not enough; we must apply.*
> *Willing is not enough; we must do."*

-Goethe.

Setting Budgetary Goals

Used correctly, a budget doesn't restrict you; it **empowers** you. You're going to establish a budget because you have financial goals that are not being met. For example, you may want to:

- Be able to pay all your bills from your paycheck—and maybe have a little left over
- Buy your first house
- Save for retirement but can't seem to find any extra money to get started
- Pay off all your credit cards and never get into debt again
- Give more money to your church or to other nonprofits
- Be your own boss
- Take a vacation
- Stop hearing from the hospital about your medical bills
- Buy a new—or at least newer—car
- Stay home with your baby
- Remodel parts of your house
- Pay for laser eye surgery
- Finance at least part of your child's college education
- Buy health insurance
- Rebuild your credit
- Find a way to care for your aging parents
- Finally build your dream house
- Take a leave of absence from your job to work in the Peace Corps
- Go back to school and begin a new career
- Buy the downtown coffee shop when the current owners retire
- Get a whole new wardrobe

Are any of these your goals? If so, budgeting will get you there, even if the odds seem impossible right now. Even if you're stuck in a job you don't like, desperately want to go back to school, have to take care of an aging parent, and have $19,000 in credit card debt, you can meet your financial goals—just as others have done before you. With a good budget, a little patience, and a whole lot of determination, you'll definitely get there.

The majority of people in today's society struggle to save. We all seem to want the newest and the best, even if we can't afford it. Credit cards and loans are leading people to financial ruin, and, even worse, schools aren't teaching us how to save properly, either.

This is precisely why the concept of budgeting is very important.

This book contains realistic and practical ways to save money and spend money wisely and how you can easily begin having more in your account at the end of each week.

What you will learn in this Book:

- How to analyze your current situation honestly.
- How to set realistic and attainable financial goals that you can actually accomplish.
- How to put together a simple budget that will help you reach your financial goals easily.
- Helpful tips and techniques that will help you successfully reduce your expenses and to spend money wisely.
- Practical and easy ways to save more of your hard-earned cash.
- Simplest ways to save money and further tips on cutting costs efficiently.
- Usual culprits for not being able to save and change your attitude about money in the process.

- How to pay off debt and start saving for retirement
- And much more...

I promise you that if you do everything in this book and stick to the plan, you'll pay off your debt and reach your financial goals faster than you ever thought possible. Even if you only use part of the steps outlined in this book, you'll be much better with controlling your finances and create a much more abundant and happier life. You will become a much more secured and confident person and much closer to financial freedom. These habits made me a totally different person. It made me become more responsible, disciplined, relaxed, secured and a happier person than ever and allowed me to really enjoy my life.

Every small action counts and compounds. The Earlier you Start, the Better.

So start right now! Just do it!

1

WHAT DO YOU REALLY WANT?

Why do you want to budget and save money? What is your dream and what is your purpose? What do you want to accomplish and what are your financial goals?

These are very important questions, and the very first questions to ask yourself.

Your dream and your WHY needs to be bigger than your excuses.

We need to change our mindset to achieve our goals.

You can accomplish your dream with budgeting and saving.

Once you discover this concept, you can be financially free. And this freedom will allow you to be creative and allow you to do things you love to do.

You can do it! Believe in yourself!

"Decide What you want.
Decide what you are prepared to give up to get it.
Set your mind on it.

Get on with the work."

-H.L. Hunt

Why Do We End Up Wasting Money?

It's normal to spend money on a daily basis, but what's not normal is not noticing that some of these expenses are actually more impulsive than necessary.

Nobody really likes to waste money. If given a clean chance to, you would really like to save or set aside money for more important things. What people have trouble with is finding that "clean chance" to start saving. Oftentimes, there are small windows opening, giving way to a chance to save up. However, 90% of the time, people fail to take it. Why do people fail to take the small and frequent chances to save up?

6 Excuses for Not Saving

1. Believing there's always another time for saving

It's common for people to put off saving because they are either already in the middle of a financial crisis or simply caught up in a shopping spree. It could also be said that people don't feel financially secure, so they think that starting to save might just make things difficult. What happens is the development of this mindset of "I can start later." Sometimes, this "later" never comes, or, when it does, there are emergencies and other more pressing financial matters that need attention. In the end, there's really no better time to start saving than now.

2. Letting money run its own course

Without organized finances and set goals, people end up buying things they want randomly and impulsively. Oftentimes, because people put off saving, they are unable to produce money to buy whatever they set out to purchase. It takes months, if not weeks, for them to produce the right amount to afford what they need. Sometimes, their goals are pushed far from their expected deadline, or, on some occasions, even entirely forgotten.

3. Thinking independence/ not being tied down doesn't require saving

People with families are not the only ones who should think about saving money. At some point in your life, you will grow old, become obsolete in your career, or simply experience financial difficulties. During these trying times, you will wish you had some "backup" stored up somewhere to help you survive. Aside from that, just because you are flying solo doesn't mean that the things you buy impulsively don't add up to much.

It might surprise you, but most single (not in a relationship) individuals spend twice as much as, if not equivalent to, a person who has children or a family. Weekly take-out dinners, designer clothes, exclusive DVD sets, and entertainment systems are infinitely more expensive than a month's worth of diapers.

4. Saving and being frugal doesn't allow you to "live" and makes you "cheap"

Being frugal is not being cheap. A frugal person saves money, values time, and happily spends money on things of value. They look for value. They need less to appreciate life and are infinitely happier than the lavish spender. They are more fulfilled, have more things that matter, and acquire less clutter.

5. Saving is too hard.

It's hard to save and to be frugal, but if you make a habit of not spending on insignificant items, you will have the luxury of being able to do many things that are not available to those at similar stages of life. If you are able to save and invest well, you will have many options available to you, which you will be very thankful for. You will be able to retire earlier than many of your friends.

Keep trying. Try a little bit every day. You will get better at spending less and saving more every day.

6. You can increase income, then save

You can definitely make more money by increasing income, but immediately as you increase your income, you will increase your lifestyle if you don't have a budget you establish to reach your financial goals. You will want to reward yourself with a new car or live in a bigger house, etc.

People who make a lot of money can also worry about retirement if they don't know how to budget and are not disciplined to take control of their finances.

2

INTRODUCTION TO BUDGETING

In this chapter, let us look at what budgeting is and why we should regard budgeting seriously.

What is budgeting?

A budget is nothing but an estimate of one's income and expenses in a given period. In other words, budgeting is a tool used to manage your finances and to achieve your financial goals. This may be both short term as well as long term in nature. It is actually simple to learn and implement but can have a tremendous impact on your life. Let us look at the two fundamental components of a budget.

Income:

Identify the different sources from which you earn every month. Most of us earn incomes from more than one source. These sources include salaries, wages, profits earned from business, investment income, rental income, etc. When you make a list, at the beginning of a month, of the amount you would be earning from the various sources, you will have a clear idea of the quantum of money you have in hand for all your expenses for the month.

Expenses:

We incur expenses for various reasons in a given month. When you write down all the expected expenses, you will be prepared to face them. Never disregard any expense as trivial. If you make several trivial payments in a month, the aggregate will be a tidy sum. Make sure you note all of them down. Some of the categories of expenses that we incur on a monthly basis are as follows:

- **Household bills:** This comprises your mortgage, electricity, water, rent and grocery expenses.
- **Financial products:** Expenses pertaining to financial products are nothing but the cost involved pertaining to your investments, such as brokerage, commission fees, etc. It also includes the premium that you pay on your insurance and medical plans every month.
- **Family and friends:** These include the expenses incurred in connection with your family and friends. At the beginning of every month, list the upcoming anniversaries, birthdays, baby showers, etc. This way you will know how much to set aside for these events.
- **Travel:** Travel expenses include the amount of money you spend on gas every month. In case you don't own a car, this will include the amount you spend on bus rides, metro, and cab rides in a month. Apart from this, if you have planned to leave town sometime during the month, then make note of the expenses that you might have to incur from the travel to and fro.
- **Leisure:** The money you spend on entertainment (such as movies or theatre), dinners, and other expenses incurred in connection with leisure activities.
- **Miscellaneous expenses:** If you incur any expenses that do not fall under the above heads, they can be clubbed under miscellaneous expenses.

■ **Contingent expenses:** The reason why we struggle a lot financially is because we do not take into consideration the contingency expenses that may come our way. Contingent expenses are nothing but the expenses that you anticipate in a given month. Learn to set aside a portion of your income every month to meet your contingent expenses.

Since these are more anticipatory than actual in nature, you can carry forward this contingent fund to the next month if it was not used up this month. This way, you will be prepared to face any emergency expenses that might come your way.

The 10 Benefits of Budgeting

We often neglect budgeting, as we regard it as a tedious task. However, this will change once you know the different benefits of budgeting. The following are some of the important benefits of having a budget in hand and following it religiously.

1. Control over your Money

We often run behind money all the time and end up being controlled by money. However, the opposite happens when you have a budget in hand. When you have a budget, you end up controlling your money. You will know exactly how much you earn and spend. This information can save you a lot of stress. For instance, when you have a clear idea of your income and expenses, you will know the amount left for repaying if you have intentions of borrowing. Often, we borrow money keeping only our incomes in mind as opposed to taking a holistic view. This will help us understand our repaying capacity and handle debt in an organized way. Apart from the control over your money, you will also be prepared even for the worst-case scenario.

on your Money goals

All of us have different financial goals. For some of us, our financial goals are aimed at settling our debts. For others, it is about increasing our assets. For others, it is about increasing the quantum of savings. Other financial goals include preparing for retirement, increasing income from the different sources, having a prosperous investment portfolio, etc. Budgeting helps you to attain your financial goals easily. This is because, with a budget in hand, you earn and spend with the desired goal in mind. With this focus, you will be able to achieve your financial goals sooner than you think.

3. Makes you Aware of what is Going on with Your Money

We often wonder what happened to all the money we earned when we become broke in the middle of the month. We keep thinking about where all the money went. Say no to these shocking epiphanies by formulating a budget. When you have a budget in hand, you will be able to keep track of what comes in and what goes out. We will also be able to track those expenses that we had failed to include in the first place and include it in the subsequent budgets.

4. Organizing your Expenditure and Savings

When we actually analyze the amount that we spend for different purposes, we will realize that a major portion of it is spent for useless reasons. We have the habit of spending money on things that we don't really need for various reasons. How often have we purchased so many things because they were cheap or bought a piece of furniture because our neighbor bought it even if we didn't need it? Since we don't keep tabs on our expenses, we don't realize how much of our money gets spent on redundant stuff. A budget helps us to identify these unnecessary expenses and helps us to cut down on them.

When we spend judiciously, we are left with more money to save. One of the hindering factors to increasing savings is the never ending list of expenses. With organized expenditure, your quality and quantum of savings will also improve.

5. Contingency Expenditure

As mentioned earlier, one of the important reasons for financial stress is that we don't plan for unforeseen expenses. With a budget in hand, we are prepared to meet even the unforeseen expenses if we set aside a sum every month on a contingent basis. This fund could reduce your stress drastically should you run into any emergency.

6. Identifying potential commitments

Although we begin preparing a budget for the month, we soon begin to prepare budgets for the long term as well. When you develop this habit, you will realize how a certain expense in a given month is impacting your expenses after six months. For instance, if you are intending to purchase a car this month, then you can expect to incur expenses for gas. Similarly, it helps you in preparing for a future commitment, such as the repayment of a debt which is due after six months. We generally tend to work on such a repayment only when the deadline nears. When you have a budget in hand, you start saving for this debt from tomorrow as opposed to after five months. This way, this repayment will not burn a hole in your pocket.

7. Increase your Income

A budget helps not only in managing your expenses but also in increasing your income. When we take note of the amount left after taking care of all our expenses, we can think of ways to use this money to increase our income. One way to increase our income would be to save the money and earn interest on it. An alternate solution would be to invest the money on shares or other securities and earn dividends on them. This way, you can multiply your existing income with less effort.

8. Determines your Debt appetite

As mentioned earlier, we tend to borrow money without taking into consideration our monthly commitments. This is why debts intimidate us. When you prepare a budget, you will know what portion of your income is left after taking care of all expenses that can be used for repayment. Based on this, you can borrow accordingly, as you will be able to commit to a fixed sum every month towards the repayment.

9. Debt management

When we have several debts, a budget helps us to prioritize them and repay them within the stipulated time. Our debts can be classified into two categories; namely, priority debts and non-priority debts.

As the name suggests, the consequence of failure of repayment of priority debts will have serious repercussions, which makes it prudent to repay them immediately. List their deadlines and apportion a sum towards their repayment every month when you formulate your budget. In other words, a sound budget will help you deal with your debts effectively.

10. Peaceful Retirement

As mentioned earlier, budgets can be prepared for long term as well. Preparing a budget for your retirement years will help you to identify what will be the income you will be earning after retirement and the expenses that you might incur, after taking into consideration the rates of inflation. When you prepare a budget for your retirement, you will know how much you will need after retirement to lead a hassle-free life. You can start saving for it accordingly, starting from today. That way, your life after retirement is taken care of.

These are some of the various benefits of budgeting. As you might have noticed, effective and consistent budgeting will help increase your savings, improve your investment portfolio, manage your debts effectively, and prepare you for your peaceful retirement.

3

SETTING YOUR GOALS

Budgeting is all about getting from where you are financially to where you want to be. And in order to do that, you've got to decide exactly where you'd like to end up. One of the reasons people often have trouble budgeting is that they haven't really sat down and thought realistically about the kind of life they want and how they might pay for it while setting the exact target date.

You're going to do things differently. You're going to start by asking yourself some hard questions.

What Do You Want?

Meet Billie DeSantos, age thirty-eight, whose budget we're going to peek at to see how this process works. Billie has worked at the same company for eight years, having worked up to management level last year. She bought a condominium six years ago, has a car payment on a three-year-old car, owes about $2,800 in credit card debt, has some money in savings, is a single parent with two kids (ages ten and fourteen), and participates in the company's 401(k) retirement plan. Billie usually has enough money to pay the bills every two weeks, although the kids' growing expenses are starting to pressure the family's income.

Billie's decided that she needs a realistic, achievable budget with some goals. **To that end, she's come up with the following list:**

- Help the kids pay for college
- Pay off the credit card
- Retire
- Put away some income in a savings account

Notice that these are all pretty general. That's fine at this stage. You can afford to be general; you're just trying to get an idea of what you'd like your life to look like. Billie wants a life in which her kids are in college (or have graduated), she's largely debt-free, and she can retire with some money to supplement her Social Security payments.

What's Realistic

Having set general goals for herself, Billie has to look at them again, this time with an eye to what's realistic and reasonable. Obviously, we'd all like to retire immediately and live in a beach house in Tahiti, but that's not going to happen. Being realistic about her goals doesn't mean Billie has to give up on them; she just has to add a time frame and some numbers to them. This is what she comes up with:

- Help the kids pay for college. Pay for half the expenses at one of the three large state universities (currently $14,500 per year for tuition, fees, room, and board) or put that same amount toward a private or out-of-state college.
- Pay off the credit card in nine months. Get the balance to zero, and then, if it's used at all, pay it off in full every month.
- Retire from the company at age fifty (in twelve years). Billie's current salary is $49,248 after taxes but before deductions for insurance and 401(k) contributions.

- Save six months of income over the next twelve years. This money would be for emergencies only, not to be touched for any other expenses.

Let's go through these adjusted goals in a bit more detail.

Help the kids pay for college

It would be great to pay the kids' entire education costs so they didn't have to take on student debt. But Billie knows she can't afford that. Instead, she creates two alternative plans: one involving in-state tuition (on which she'll receive a discount); the other for a more expensive alternative but one that will mean the students taking on more debt. When we come to consider budgeting for college, we'll see there are some other alternatives as well as the ones Billie's come up with.

Pay off the credit card in nine months

Currently, Billie's credit card debt isn't too high. The important thing, though, is that she wants to get out from under the constant interest payments. Again, she's realistic enough to know she can't simply pay the $2,800 she owes in one lump sum; it's going to take some small payments spread over most of a year. But the important thing is she's got a plan.

Retire from the company at age fifty

Retiring at age fifty is probably a bit unrealistic given her salary and age, although this is a goal she can adjust. Retiring at fifty means she has twelve and a half years before she can start receiving Social Security benefits, so this might be a bit tight. Still, it's a good place to start from.

Save six months of income over the next twelve years

At Billie's current rate of pay, this would amount to $24,624 after taxes and before deductions. This is a reasonable sum to set aside for emergencies such as medical problems, accidents, or other unexpected events.

Stretch Goals

It's a good idea to have a stretch goal—something you aspire to but only can attain by working very hard. Billie has such an aspiration: When she retires, she wants to open a bed and breakfast in a small coastal town. B&Bs in similar towns currently cost about $650,000 for the building and operation, but that price will surely rise in the next twelve years. On the other hand, if Billie is successful, the B&B will also provide a source of income during her retirement.

Income and Expenses

Now let's see just how realistic these goals are, given Billie's current income and spending patterns.

Billie's biweekly income after taxes (which provides her with a refund of about $450 per year), company-sponsored health and dental insurance (at a cost of $55 per pay period), company-sponsored life insurance for $250,000 of coverage ($30 per pay period), and 401(k) contributions ($75 per pay period, matched by the company) is $1,892, which totals $4,100 per month. Billie also has $1,700 in savings.

Her monthly expenses are as follows:

Mortgage on the condo (30 years at 7.85%) includes taxes and insurance	$1,492
Car payment (4 years at 5.9%)	$342
Utilities	$375
Food (including eating out)	$675
Toiletries/haircuts	$85
Spending money/allowances	$200
Car maintenance/insurance/expenses ($1,600/year)	$133
Vacations ($2,800/year)	$233
Clothing ($3,200/year)	$267
Gifts and contributions	$200
Credit card debt ($2,800)	$50
TOTAL	$4,052

Ways to Reduce Debt

Billie's monthly obligations just about equal her monthly income, so to achieve her financial goals, she'll have to eliminate some expenses. Here's what Billie decides to do:

- **Keep the car and car payment.** After paying off the car in one year, continue to drive it for five years after that, putting $342 into savings each month for the next car. No monthly savings.

- **Cut down on utilities.** Get rid of her landline (go cell-only) and install a programmable thermostat (at a cost of $46) to save on gas bill. Estimated monthly savings: $83.
- **Spend a maximum of $125 per week on groceries.** Limit eating out to pizza or Thai takeout once a week. Estimated monthly savings: $175.
- **Eliminate the small stuff.** Keep Starbucks visits to once per week, borrow magazines and DVDs from the library, and otherwise reduce monthly spending money to $150 ($100 for both kids' allowances; $50 for Billie). Monthly savings: $50.
- **Investigate car insurance options.** Lower annual insurance costs by $400. Monthly savings: $33.
- **Limit vacation spending.** Reduce annual amount to $500 per year by being creative. Monthly savings: $191.
- **Allow each member of the family $600 per year to spend on clothing and shoes** (teaching the kids budgeting skills in the process). Any more than that, the kids will have to use their allowances or get part-time jobs. Monthly savings: $117.
- **TOTAL Monthly Savings: $649.**

Meeting Goals

To meet her financial goals, Billie must increase monthly savings and investments:

- **Refinance the mortgage** on the condo at 5.8% for fifteen years, paying it off in twelve (so that it can be sold, debt-free, to help pay for the B&B, which will then be mortgaged for fifteen years). Monthly increase: $290.
- **Use savings** plus increase in monthly payment to pay off credit card in nine months. Monthly increase: $122.
- **Begin saving for college** in a 529 plan. Monthly new expense: $600.

- Save six months of income over the next twelve years. Monthly new expense: $171.
- TOTAL monthly increase: **$1,183**

Revisiting the Sample Goals and Priorities

Billie is $534 short each month, so it's time to revisit the listed goals to see which can be changed or eliminated. Here's the revised list of goals (changes in italics):

1. Help the kids pay for college.
2. Pay off the credit card in three months, and begin saving for the kids' college fund only when it's paid off.
3. Retire from the company at age fifty-four (in sixteen years) and open a bed and breakfast in a small coastal town.
4. Put away six months of income in a savings account over the next sixteen years.

These changes mean the following financial adjustments:

- **Refinance the mortgage** on the condo at 6 percent for thirty years, with the understanding that in eight years (when college savings will no longer be necessary), the money currently used to save for college will be redirected to the mortgage. Making those large extra payments toward the mortgage after the kids finish college will result in the mortgage being paid off in eighteen years, not thirty. Reduces monthly shortfall by $310.
- **Put away $130 per month into savings** (instead of $171) over the next eight years, and then increase savings with reduction in food, utilities, clothing, and other expenses because the kids will have left home. Reduces monthly shortfall by $50.
- **Delay contributions to 529 plan by three months,** using that money plus funds from savings to pay off credit card debt. After

credit card is paid off, begin saving for college in a 529 plan, putting $418 (instead of $600) away, with the understanding that all future promotions, raises, and tax refunds for the next eight years will go directly to the college savings account. Reduces monthly shortfall by $304.

■ TOTAL monthly increase from current spending: $0

Billie has created a working budget. It won't be easy to cut back, but the family does still have some discretionary spending money, the kids' educational savings are in good shape, and Billie will realize the dream of owning a B&B in just sixteen years.

Billie's success depended on her ability to do two important things that are essential to the budgeting process: prioritize and compromise. She had to prioritize her goals and decide how much she wanted to achieve each of them. She had to balance long-term goals against short-term ones and decide which were immediately realizable and which were going to take longer. And she had to figure out what she'd be willing to give up in order to realize those goals. This process of prioritizing and compromising is at the heart of good budgeting.

Now that you see the basics of how it's done, let's roll up our sleeves and begin the work of evaluating what you have, what you're spending, what you owe, and how you're going to create a plan to realize your dreams.

4

WHAT DO YOU HAVE?

Before you can create a budget, you have to know every detail of your financial situation. Although you probably understand in general how much you spend and where you spend it, you may be amazed at how much you actually spend on certain items that don't seem like they could add up so fast.

The process we're going to undergo in the next chapters is designed to give you an unflinchingly honest appraisal of where you stand. If you have a tendency to become overwhelmed easily, keep a friend on standby whom you can phone for support, and keep upbeat music playing in the background as you put together your income and expenses.

What Counts as an Asset

We're going to start by looking objectively and calmly at what you have —that is, your assets. An asset is essentially anything that you own. That can include regular income (such as your paycheck); occasional income (such as an inheritance or a tax windfall); your home, if you own it; your car, assuming you own one; and all other material goods that you own. We're not proposing that you sell your home or your car—don't panic— it's just that it's necessary to know your total worth in order to figure out how to realize your dreams.

What doesn't count as an asset is money you can't count on. Getting lucky in the lottery this week doesn't count as income because you could just as easily get nothing and be out the price of a lottery ticket. The same thing is true of things you will or might own in the future. If an elderly relative has said that when she passes she's going to leave you her big house in the country, that's nice, but it still doesn't count as an asset now.

Some people argue about whether money you're owed should count as an asset. I'm inclined to think not, because you have no guarantee of its being paid back. When the money's actually in your bank account, I think you can count it as an asset, but not until then.

Determining Your Income

We'll concentrate first on your income because that's the most immediate and obvious asset and the one that has the most immediate impact on your budget. Your income includes any money that comes into your possession and can be counted on in the near future. Your paycheck is considered income, but income isn't limited to a paycheck you receive from your employer—it is also a disability payment, a welfare check, a Social Security check, alimony, child support, self-employment income from a small business, and so on. Whatever money comes in—money that you can count on—is what you want to consider as income.

Determine How Often You're Paid

First, determine how often you're paid at work:

- **Weekly:** Common for temporary and contract work
- **Biweekly:** The most common way companies pay their employees—usually every other Friday
- **Semimonthly:** Often paid on the first and fifteenth of each month
- **Monthly:** One paycheck each month
- **Quarterly:** Four times a year—this is rare
- **Semiannually:** Twice a year—this is also rare
- **Annually:** Almost unheard of unless you're on a board of directors

If you're paid on commission and aren't exactly sure when your next check will be coming, review your income from last year and use that as a starting point. If, however, something has changed since last year that may cause you to make fewer sales this year, adjust accordingly.

If you get paid monthly (or even less frequently), you may have a harder time than most with your budget. The amount of your check may seem like a lot at the beginning of the month, but three or four weeks later, your expenses may have exceeded that check. A strict weekly budget can really help.

Identifying Your Sources of Income

Worksheet 4-1 helps you identify all your sources of income, add them up and figure them on an annual basis. In this worksheet, you calculate all of your income for a given pay period and multiply it to get an annual amount. Be sure to write down the net amount of each paycheck—that's the amount you take home after the taxes, insurance, union dues, and other items are deducted.

Worksheet 4-1: Your Income

Sources of income	Amount	Multiply by	Annual amount

Remember that this is not projected income. Don't write down the income you think you'll have after a promotion or other situation; what you want to look at is exactly how much money you have to work with right now.

Your House as an Asset

Now that we've gotten a good picture of your income, let's consider some of your other assets. We'll start with your house, which is probably the most valuable thing you own. Note that this only applies if you own your house or condo. If you rent or have some other living situation, you can skip the part of this chapter that follows.

Later in this book, we'll discuss refinancing, which is the major way to save money in regards to your house. For right now, let's just discuss the basics.

Understanding Equity

Equity is the portion of your house that you own, mortgage free. You can calculate your equity as follows:

1. Determine the current value of your home. This amount may be higher, and in some cases much higher, than the amount you paid for it. The value may also be lower than what you paid if the house was overvalued when you bought it or if the real estate market in your area has slumped. A mortgage company requires an appraisal, done by a professional, to determine this value, but you can guess based on what homes in your neighborhood have been selling for. Check realtor sites and the newspaper to get a list of houses in your area and their asking prices. Then make an equivalency to your own house, using houses that have comparable square footage, number of bedrooms, and lot size. You don't have to be exact; this is to give you a ballpark figure so you know what you're working with.
2. Determine the current payoff on your mortgage. If you don't receive a monthly statement or receipt that tells you the payoff amount, call your mortgage company and ask for it.
3. Subtract the payoff from the current value. This is the equity in your home.

Instead of calculating the current value of your home, some lenders use the value from when you bought the home. If that was more than a couple of years ago, the current value may be much higher (although it also may be lower, depending on how housing prices in your neighborhood have performed).

Now let's move on to your car, often the second most valuable asset you possess. If you own a car, go online to look up its blue book value. Remember that cars depreciate swiftly, so don't be surprised if today it's worth a lot less than what you paid for it. Some factors that will affect its value include its condition, the mileage on it, and any "extras" such as a sun roof, advanced sound system, GPS, and so forth.

Finally, add in any extra assets, including property you own and anything of special value, such as antiques or stamps. For right now, we're going to concentrate on income because that's the part of your assets that most affects budgeting. But keep your valuation of your other assets handy because they represent elements in your financial picture.

Now you've found out how much income you have and what assets you own, including the value of your home and how much equity you have in it and the value of your car. It's time to take the next step: investigating how much you spend and what you spend it on.

5

WHAT ARE YOU SPENDING?

Income and assets are only half the equation in your budget. Now you need to tackle your spending. A word of caution here: Some people are tempted, when creating a budget, to fudge their spending numbers a bit. It can be embarrassing when you realize how much you're spending every week eating out or going to movies. But it's absolutely essential in the following worksheets that you are completely honest with yourself about how much you spend and what you spend it on. Don't worry about putting it down in black and white. You don't have to show it to other people if you don't want to. You need it in order to make a careful and honest evaluation of your outgoing funds. If you don't like where you're spending money...well, I can show you how to fix that.

You have two ways to free up money for your financial goals: making more or spending less. Neither one is better than the other, right? Wrong! If 25 percent of your income goes to state and federal taxes, then for every extra $1 you earn, you can use only $0.75 to pay off debt or save for the future. But if you can save $1 of your expenses, you can apply all of it to your debt or put it into savings or investments. In other words, when you cut expenses, you are saving after-tax dollars. $1 saved is now $1.25 earned (depending on your tax bracket).

What Are You Spending Every Day?

The method of totaling your expenses on Worksheet 5-1 is simple: You either report what you spent last week—day by day, expense by expense—or you start fresh this week and record every expenditure going forward. Use one worksheet for every day. If you record your expenses this coming week, be sure you don't try to be "good" and spend less than you usually do.

Worksheet 5-1: Daily Expense Sheet

DAY:			
Item	Amount	Item	Amount
	$		$
	$		$
	$		$
	$		$
	$		$
	$		$
	$		$
	$		$
TOTAL:	$		$

Categorizing and Prioritizing Your Daily Expenses

Now, review your daily lists and categorize them in the most logical way you can: coffee, breakfasts, lunches, clothing, toiletries, groceries, and so on. Use Worksheet 5-2 to record your findings. Ignore the "Priority" column until you've listed all of your expenses by category.

Worksheet 5-2: Daily Expense Summary

Category	Total Amount	Priority (1 - 5)
	$	
	$	
	$	
	$	
	$	
	$	
	$	
	$	
	$	
	$	
TOTAL:	$	

Now go back and assign a priority to each category—5 being something you absolutely can't live without and 1 meaning you'd barely notice if you no longer spent money on this one. Your priorities don't necessarily mean that you will continue to spend this money in the exact same way. They're just a way to understand your spending habits.

What Are Your Monthly Expenses?

Having looked at your weekly spending patterns, let's move on to those expenses that occur on a monthly basis. Recording your monthly expenses in Worksheet 5-3 works just like your daily ones, except that while daily expenses are often cash expenditures that you may not really notice, monthly expenses such as rent and utilities are more likely to be paid by check or electronic transfer. For these monthly worksheets, you'll want to go back through your receipts, checkbook register, and bank statements, and also use your memory. If, on the other hand, you simply record all your monthly expenses starting in January, you'll understand your January expenses on January 31, but you won't have a clear picture of your December spending until a year from now, and that's valuable time that you could use to reach your financial goals instead of getting deeper into debt. Instead, to more quickly get a clear picture of your monthly expenses, dig out your bank statements, checkbook register, receipts, and so on. If you use online banking, go back through the electronic register of your transactions.

Be sure not to double up on daily and monthly expenses. If you've already recorded a certain expense on your daily expense sheets, do not record it here. Use one copy of Worksheet 5-3 for each month.

Worksheet 5-3: Monthly Expense Sheet

MONTH:		
Date	Item	Amount
		$
		$
		$
		$

		$
		$
		$
		$
		$
		$
		$
		$
		$
		$
		$
		$
		$
		$
		$
		$
TOTAL:		$

Categorizing and Prioritizing Your Monthly Expenses

Now you're going to group your monthly expenses into categories (utilities, rent, insurance, etc.) and add them to Worksheet 5-2. Then prioritize those categories. Sounds familiar, right? It's just what you did for your daily expenses.

Prioritizing Which Items You Want to Spend Money On

You should now have a list of your expenses by category, with a priority attached to each one. If you have enough income to reach all of your financial goals and still spend money the way you currently do, you won't need this prioritized list. However, it's very likely that you can't meet your financial goals if you continue to spend in this pattern. (Remember Billie's experience outlined in Chapter 3.) If that's the case, use this list to choose the areas that you absolutely do not want to cut back on (these are the items that have a priority rating of 5). If you have too many items with a high priority to meet your financial goals, sub-prioritize those items so that you come up with just a few. Spending on these items will make all the cutbacks easier to swallow.

Remember that you are the only one who can determine your priorities. If you would rather drive an old car so that you can still afford to buy organic fruits and vegetables, do it. Remember also, while you're making these hard choices, that your goal here isn't to make your life less enjoyable. Instead, it's to find a path to realizing your dreams. Such a path may require you to make some sacrifices in the present in order to reach your goals in the future.

Occasional Splurges

So does keeping to a budget mean you can never go out for dinner again or spend a romantic weekend in a nice resort hotel? No, of course not. It just means that you need to budget for these things.

When you're tracking your expenses, look for these kinds of "splurge" items. How much have you been spending on them? Assuming you want to continue to do so, look for ways to splurge but without necessarily spending as much money.

There's nothing wrong with spending money to enjoy yourself— provided that this is what you want to do and that you've included it in your budget.

Investing in YOURSELF

So what should you spend money on? One of the first things that you should invest in is yourself, because being able to save money starts with you. The better health you are in, physically and mentally, the better you can perform at anything! Besides health, you should also make it a priority to live in a good environment, have strong relationships, and have the willpower to get into some great habits, and that includes spending habits! You will learn more on how to sharpen your spending habits later on, but for now, I will share with you the importance of investing in self-improvement.

Self-Improvement

I believe that you can never improve yourself enough. There is always room to learn something new and there is always a way to get better at something you're really good at. When it comes to spending money, it is important to pay your bills, but the next thing that you should focus on is putting money towards self-improvement. This includes your physical health, mental health, and education.

Physical Health

The healthier you are physically, the less likely you are to develop illnesses and ailments, and you'll also have a higher chance of living a longer, happier life. In a way, money can't buy that feeling of success, but if you pay attention to what you do put your money towards, your chances of reaching financial freedom can be much greater.

- First thing's first. It is important to eat a healthy, balanced diet. Eating out sometimes as a treat or for a special occasion is okay, but if you do your own grocery shopping and learn how to cook, you can save money while spending money and improve your physical health all at the same time. Pay special attention to what you are buying and focus on getting plenty of fruits, vegetables, whole grains, and protein.

- Secondly, invest some time and money into exercise. You can purchase your own home gym equipment or get a gym membership. If you're not looking to invest money into exercise, you can simply exercise outdoors. Just 20-30 minutes of walking a day is great for you! Adding exercise to eating a healthy diet is key for maintaining a healthy weight and keeping your immune system in tip-top shape.

- Third, get plenty of sleep! If you do not get enough sleep for your age, your body will not be able to function as well. The less productive you are, the less likely you are to accomplish successful things. Invest in some good shut-eye time and your body and wallet will thank you!

- Make regular doctor's appointments to make sure that you are doing everything correctly. Your doctor can help you figure out if your body has any deficiencies and if you should make any changes to your eating and exercise plan based off your individual health situation. Some people have vitamin/mineral deficiencies and will need to take supplements.

Mental Health

Your mental health is just as important as your physical health. Investing in this area should also be a high priority. Mental stress and depression can make you care less about your spending willpower and you are more likely to feel horrible about yourself overall. Here are some great ways to invest in your mental health, and some of these strategies won't even cost you anything but time.

■ **Practice yoga**. Yoga can be a very relaxing type of exercise and you can do it by yourself or with a group. Scientifically, yoga manipulates your body's stress response system, meaning that it can help offset anxiety-related issues, such as increased heartbeat, and can help make you less sensitive to pain. A wise investment would be to take a class or buy a DVD that teaches you the basics of yoga.

■ **Meditate**. People have used meditation as an anti-stress technique for centuries. It can help you get in touch with your body, your thoughts, nature, or anything else. It is a very relaxing technique that you can perform in any dark, quiet room. Best of all, the only resource you need to invest into meditation is time.

■ **Try Aromatherapy**. Aromatherapy can be very relaxing for some people, as certain scents can help your body release stress and anxiety. A wise investment would be to buy a small oil diffuser and try a few different scented oils to combat stress. I have found that lavender, patchouli, valerian root, chamomile, and eucalyptus oils work best for anti-anxiety and relaxation.

Environment and Relationships

Another area of self-improvement that you should invest your resources into is your environment and relationships. You can be in the best health ever, but if your environment and relationships are not in tune with it, then you probably won't get very far.

As far as your environment goes, one of the most important things you can do is keep it clean. Clutter is known to actually cause more stress, so if you live in an organized place, you are less likely to be stressed out and more likely to be productive. Invest some time into cleaning up and you'll find yourself feeling great! Do your best to organize everything into neat little areas that are easy to access and well-organized. If you live in a nice outdoors area, take advantage of that and go for strolls in your local parks. If you live in a more industrial area, such as a city, go out, get involved, and make the most out of your life and living arrangement. The friendlier you are with people, the more likely you are to make friends. You can even increase your chances of getting great job opportunities just by talking to people.

That brings us to the topic of relationships. If you are missing great, strong, healthy relationships in your life, you're risking your physical and mental health. The key is to surround yourself with positive, supportive, and like-minded people. Negative people will only bring you down. Invest some time into developing good relationships with your friends and loved ones.

The stronger they are, the stronger you can be.

Education

Finally, one of the best things that you can do to improve your life is to invest in your own education. Whether you spend $100,000 to go to a well-known university (we'll get to overcoming your debt problems later), or just $10-25 on books and materials each month for personal education, you can't go wrong. I know a lot of people who have been out of school for years yet they continue to read and learn because it fulfills them and it helps them become better people. I think it's very easy and wise to make monetary investments into your overall self-improvement because you could gain something truly valuable to apply to your life that will pay huge dividends for you throughout the rest of your life. The more you learn, the more you will find doors of opportunity opening for you. Learning new things can help you feel confident, inspired, and motivated to do great things with your life. Also, the more you learn, the more job opportunities you can get along with great ideas that could earn you or save you lots of money.

The point of this chapter is that investing in yourself is crucial for being successful at anything. All of the points listed in this chapter can easily make up the foundation of a great, successful life. By investing your time and money into these things, your chance of a higher return in your life can increase. For example, if you spend $10 a month on general self-improvement books, you may find that it's easier to talk to others, be confident, and be motivated, which may lead to a promotion at your job or the courage to follow a lifelong dream.

6

THE BEST WAYS TO SAVE

Build a Savings Account

The two most common types of checking accounts that most people have are checking and savings accounts. By keeping money in a checking account, you can gain easy access to that money for spending by writing out a check or swiping a debit card. A savings account is a little different. Many people put money into a savings account for the purposes of, obviously, saving. You might open a savings account to save money for emergencies, retirement, or a down payment for a large asset. Money that you put into a savings account acquires interest over a period of time. Best of all, you don't need a large amount of money to start a savings account. Depending on your bank, you might only need as much as $25 to begin. Some banks charge a low monthly fee or offer them for free to open a savings account and interest rates will vary by bank. Always shop around before deciding on a bank to open a savings account with. You should generally be able get one for free.

There are many benefits to opening a savings account. First and foremost, your chances of spending that money are much less than if your money was in a checking account. Secondly, your money is safe in a savings account. If your house were ever to get burglarized, or if a tornado ripped through your neighborhood and swept your house away, your money would still be safe and sound in the bank. Money in a savings account is also safe because it is insured by the FDIC if you live in the USA. So if your bank were to close, you wouldn't lose your money. Finally, many people open savings accounts to accrue interest. That is when your bank pays you money to lend your money. When that happens, your bank will usually pay you interest every month.

Basic savings accounts, which usually only require small fees to get started, only earn a small amount of interest each month. A market money account gains a higher amount of interest but often has limitations. For example, you need a lot of money to put into it, and you can only make a small number of withdrawals.

Once you've opened your savings account, your bank will give you a log where you can track your money. You can also track your deposits, withdrawals, fees, and interest gains by reading your monthly statements.

The Simplest Ways to Save Money

Now that you know what to do in order to get a handle on your finances, it is time to learn about little, easy things you can do to save money. Sometimes, saving money is as easy as making a simple switch or thinking about your alternative options. You will learn some great ways to simply save money in this chapter!

Fold it, put it in your pocket, and keep it there

One of the best and most commonly quoted advice about saving money is this: "The best way to double your money is by folding it in half and putting it in your pocket." In reality, literally folding your money bill in half won't double its amount. However, the total value of what you have can and will be doubled for each day you follow this habit successfully. If, for example, each day you set aside five dollars without touching or scheming to spend it on something, by the end of the week, you would have added 35 dollars to your savings.

How It Helps:

Folding your money over or simply keeping it in your pocket won't just save you some money, it will help control those shopping impulses. One of the most common mistakes people make when they get their money on payday is splurging. It's no surprise if you can relate to a scene like this: Payday arrives and you feel as rich as a king. It only takes a week, or less, of seemingly endless wealth until you're broke or nearly broke. By the end of the month, you're sulking, cutting down expenses and waiting at the edge of your seat for the next payday. The cycle happens, and it will continue to happen over and over again until you decide to act upon it.

Saving Every Penny

Most people think of small bills as insignificant because, well, they are small or have less value and would not really be of any use in emergencies. The truth is, they are actually the most significant portion of the idea of "savings."

How It Helps:

Small amounts are what a person builds up and turns into millions. Mountains are not made of singular gigantic boulders. They are made of sand, stones, rocks, and some boulders. In the financial comparison of mountains to savings, it's the pennies that make up the base and not the hundreds. While saving pennies won't make you rich, it will change your relationship with money. If you practice putting your coins in a tin every day, you will condition yourself to save. This brings us to the next overlooked technique in saving money, for long- or short-term purposes.

Starting Small and Starting Now

Starting small is the easiest way to save, but starting small is belittled by most because of the expected value it achieves. "Every little bit counts," that's one saying that applies to a lot of things, including saving money. Since the idea of saving up is a continuous and consistent habit, even the small contributions add up to the pile. It's also easy to save by starting small because it gives people enough freedom to buy the things they need or want. It won't feel like such a responsibility, which will lengthen its chances of actually growing to be a successful habit.

Investing Your Money

Besides opening a savings account and letting the bank lend out your money, there are many other smart ways to invest your money. This section will give an overview of the different types of investing. Many people are turned away from the idea of investment because it can often be a risky venture. However, as long as you go into it with knowledge and goals, your chances of being successful can be higher.

1. Invest in Property

2. Invest in Other Businesses

* Capital Gain Incomes
* Invest in Stocks
* Invest in Bonds
* Real Estate
* Assets

7

SAVING AROUND
THE HOUSEHOLD

While you're going through the process of evaluating and re-evaluating—and re-re-evaluating—your budget, it's time to take a look around your household with an eye to saving money.

What you do in this regard is completely driven by how you balance what you're doing to save money against your short-term and long-term goals. If you decide that, even if it's expensive, you just can't give up your cable package, that's fine. It just means that cable TV is more important to you than some of your other goals. There's nothing wrong with this, and you should not judge yourself or tolerate others judging you for making decisions like that.

My goal in this chapter is not to demand that you save money in all the ways I suggest; it's to give you options that will enable you to save money if you decide to take advantage of them.

Prepare Your Own Meals

Even a generation or two ago, eating at restaurants was reserved for special occasions, and eating prepared foods was practically unheard of except among a few lifelong bachelors. Today, it's actually less common to make a meal from scratch than it is to eat at a restaurant, get takeout or fast food, or prepare a meal by mixing together ingredients from a box. Do people actually make meals from scratch?

Yes, they do. One way to save hundreds of dollars every month is by making your own food. You don't even have to be very good at it: Even the least-seasoned chef can boil pasta and mix it with tomato sauce or broil a piece of chicken or beef. If you can make toast, you can cook. And the more you cook—even the easy stuff—the better you'll get at it; then you can progress to more difficult meals. Also, you can eat a much more balanced diet.

The trick, of course, is that you have to take time out of your life to shop and cook, and most people don't have time these days. But if you're trying to meet your financial goals and don't have the opportunity to make more money, you can save a great deal of money by taking the time to cook.

Seeing How Much Eating Out Really Costs

To find out how much you can save, think about a plate of fettuccine alfredo with chicken from a local restaurant. Visualize what's in a plateful, even if you don't know much about cooking. Offhand, you might guess that there's about a quarter pound of fettuccine, half a cup of cream, two tablespoons of butter, a quarter pound of chicken, and maybe a few other ingredients, but that's close enough. List the ingredients in your favorite dish on Worksheet 7-1, and the next time you're at the store, price it out. You can make a big portion of fettuccine alfredo for about $2.90—it will cost at least $12.95, plus a tip, at your local Italian restaurant.

Worksheet 5-1: Grocery versus Restaurant Comparison

Ingredient	Price to Buy at Store
	$
	$
	$
	$
	$
	$
	$
	$
	$
	$
	$
	$
TOTAL:	$

Of course, you won't get your meal prepared and served, but you also won't have to drive to the restaurant, wait in line, or pay a tip. Want to save money? Prepare your own food!

Clean Out the Fridge

If you tend to eat out a lot (or eat in with pizza, Chinese takeout, or other fast food), you may feel guilty about the fact that you're not cooking. To ease this guilt, you may go grocery shopping every week or two and buy vegetables, meats, dairy products, and other perishable foods with the intention of changing your ways; but you don't change your ways, and the perishables go bad, and you throw them all out. A few weeks later, you start this vicious cycle of wasted money all over again.

Here's the thing: Either decide you're going to have someone else cook for you, or decide you're going to cook for yourself. Neither way is inherently good or bad, but don't do both and waste food in the process. If you buy the food, stop eating out until it's gone. If you know you're going to eat out, don't buy the food. You could save a few hundred dollars a month.

Brown-Bag It

If you go out to lunch every day—or even a few days a week—you can save quite a bit of money every month by bringing your lunch from home instead. It doesn't have to be fancy; in fact, if you make just a bit more than you need for dinner every night, you can pack the leftovers for lunch the next day. Or pack a sandwich, yogurt, and fruit—simple and cheap!

To find out how much you really spend on lunch, review your daily expense sheets in Chapter 4. Then make a sample menu for your lunches and write the ingredients in Worksheet 7-2, comparing that total to your eating-out total.

Worksheet 7-2: Brown-Bag versus Eat-Out Lunch Comparison

Ingredient	Price to Buy at Store
	$
	$
	$
	$
	$
	$
	$
	$
	$
	$
	$
	$
TOTAL:	$

Budgeting Tip: *If eating out at lunch is a big part of your social life, don't stop it completely. Instead, limit it to one day per week or a couple of days a month.*

Coupons Are Your Friend

Coupons are free money, but only if you use them for products you would have purchased anyway. Companies offer coupons because they want you to try their products, and they figure they'll give you a bonus for taking a risk, but this isn't how you save money on food. Instead, as you peruse coupons, cut out only the ones for products you already use or products that you're willing to use because you're not loyal to another brand. If, for example, you couldn't care less what kind of detergent you use, cut out all laundry coupons and use the ones that save you the most money. But if you never eat anything but Sugar Cereal for breakfast, don't cut out coupons for Fruity Cereal, no matter how much of a savings you'll get. All that will happen if you do cut out those Fruity Cereal coupons and use them is that the cereal will sit on your shelf until it goes bad. Coupons should save you money, not promote waste.

Cutting out coupons is just the first step. To actually use them, you'll have to have them handy and organized. One cheap, simple way to keep them accessible is to put them in a 3×5 card-file box, organized by category. File the coupons under their appropriate categories, placing them in order by expiration date so that the coupon that will expire first is the one you see first when you flip to that category. When you add new coupons, flip through each category to see whether you have any expired coupons that you need to toss out. Keep the box close to your keys so that you never forget to take it into the store with you. Or, if you tend to forget it, keep it in your car.

Budgeting Tip: *Look for coupons in the Sunday paper, keeping in mind that coupons usually aren't offered the weekend of, or just before, a major holiday. Also, if you're buying the paper only for the coupons, make sure you're saving more than the cost of the paper each week. Also look for electronic coupons such as Groupon. Visit the Groupon website at www.groupon.com.*

Look for Food Bargains

The Sunday newspaper can be a great source of information about the cost of products in the stores in your area. Before you peruse the sale flyers, write out your weekly or monthly shopping list. Then look through that week's advertisements, noting on your shopping list which store has the best price on the items you need. Then make a quick stop at each store to buy only the items on sale that week.

Price Matching

If you have a store in your area that price matches—and most now do—you don't have to do all that running around. You simply inform the checkout clerk that you're going to price match an item, and you get the least expensive advertised price on that item.

Buy Sale Items in Quantity

If you see a great sale price on an item that you use a lot—and if the item isn't perishable and you have the space to store it and you have enough money in this month's budget to pay for a large quantity and you're sure beyond a doubt that you will actually use this item up—buy a lot at the sale price. Suppose you make a tuna fish sandwich for lunch every day and usually pay ninety-eight cents for each can of tuna. If you see it on sale for forty-nine cents, buy as much as you can store and afford, because you know you'll use it. But if you see bananas on sale, buy your usual amount. You can't possibly use up a large amount of bananas before they go bad.

Join a Wholesale Club

One way to get sale prices every day is to shop at a wholesale club, such as Sam's Club or Costco. If you decide to go this route, make sure you're saving more money than the annual membership fee, and be sure that you don't spend more than you should in the name of "But it's such a good deal." Wholesale clubs can ruin your budget, so beware!

Budgeting Tip: Some wholesale clubs offer memberships only to employees of small businesses, schools, churches, credit unions, and other groups. They also sometimes offer memberships to family members, friends, and neighbors of members. Others are open to the public. To join, call your local wholesale club to find out the membership requirements.

If you do buy at a wholesale club, apply the same logic that you would for buying large quantities of any sale item at your regular store: Buy bulk quantities only if you have the storage space, are sure you'll use it, can keep it from spoiling, and have the money in your budget to pay for it.

How Does Your Garden Grow?

If you eat a lot of veggies, you know they can be expensive. Yet for just a few dollars for the seeds, you can grow an entire garden of fresh vegetables every year. And if you have extras of easy-to-grow vegetables like tomatoes, you can freeze them for use in pasta sauces in the winter.

Using Your Patio or Balcony

Even if you don't have an extra acre out back to grow a garden, you can still raise vegetables. Many veggies grow well in outdoor containers on a patio or balcony, if you're careful to keep them well watered, well drained, and protected from freezing weather at night.

Growing Organically

To save money and protect your health, grow your vegetables organically. The trick to gardening without chemicals is to start with excellent soil. Improving your soil may cost you some money, but it'll pay off for years to come. To be sure, however, that you're actually getting more benefit from your garden than you're paying in soil, seeds or plants, and equipment, total up what you get out of your garden the first year and compare that to what you spent to get started.

An alternative to growing your own garden is shopping at a farmer's market in your area. You'll usually pay lower prices than in a grocery store for fresher, less-processed fruits and vegetables. And you can still freeze the bounty when you find especially good deals on fresh produce.

Turn the Thermostat Down (or Up)

Now let's turn to another huge domestic expense: your heating and cooling bills. Of course, you could just not heat your house in the winter or cool it in the summer, but unless you live in a highly stable climate—say, Hawaii—that's probably not possible. A simple way to cut your heating and cooling costs is to turn your thermostat down one degree in winter and up one degree in summer. One degree—which you probably won't even notice—can save you up to a hundred dollars a year on your heating and cooling bills.

A simple way to do this is to use a programmable thermostat. These thermostats automatically turn your temperatures up and down at preset times. So if you are always in bed by 11:00 in the winter, program the thermostat to turn down the heat at 11:15, saving you money all night. It then turns the temperature back up at 6:30 in the morning, so you wake up to a toasty house. It turns the temperature down again while you're away at work and turns it up just before you get home. These thermostats are easy to program—look for one that offers daytime and nighttime settings, plus separate settings for the weekend when you're likely to be home more and sleep in later. Because programmable thermostats actually turn the temperature down, they pay for themselves in a couple of months.

Budgeting Tip: You can purchase a programmable thermostat at your local home improvement store. Be sure to get one that has both weekday and weekend settings, especially if you tend to wake up later in the morning on Saturdays and Sundays.

Get Your Books and DVDs from the Library

From books to CDs, DVDs, and audio books, your library has a wide range of free opportunities to entertain you and your family. As you probably know, books can be very expensive, especially textbooks. In most cases, your local library may have the same book that you could buy in a store. And library cards are also free. You could even stay there and read all day if you wanted. Going to the library is a fun activity, and best of all, it's much cheaper than buying new books in the store.

Getting Rid of Cable/Satellite and/or Your Landline Phone

Although you may think that cable or satellite TV is part of life's necessities, these are really just extra services that you should subscribe to only if you have plenty of extra money each month—after you pay all of your other financial obligations. It is possible to live without them, and you'll read a whole lot more if you get rid of your TV altogether.

If you're using your cell phone most of the time anyway, consider getting rid of your landline, too.

Because you're probably cutting way back on your expenses, don't overlook these simple ways to save a lot of money. Prices for these services vary greatly from one area to another, but here's an example:

- Cable: $60 per month × 12 months = $720 per year
- Landline: $65 per month × 12 months = $780 per year
- Total Annual Savings: $1,500 per year

Consider this: If you're currently trying to pay off $2,500 in credit card debt and make no other changes to your income or expenses except to get rid of your cable service and landline, you'll be debt-free in less than two years. And you'll save even more if you are paying for premium movie channels or if your landline has several premium features or doesn't include free long distance.

Okay, so perhaps you accept the idea that these services do cost money and that you'd be in better financial shape if you got rid of them. Your resistance comes from not knowing how you could possibly live without them. Here are some alternatives.

Alternatives to Cable and Satellite

Instead of subscribing to cable or satellite TV, which offer you dozens or even hundreds of channels to choose from, record every cable, network, and public television program you think you'd enjoy. Build up a collection of favorites to rewatch anytime the urge strikes. Don't forget your public library, either: Most loan DVDs of movies, documentaries, and miniseries free of charge, and some even loan popular series, too.

If you get such terrible reception in your area that you can't even watch TV without cable or satellite, you have two low-cost choices. One is to sign up for just the basic cable coverage, which gives you network stations, public television, and perhaps a few other stations. The other is to stop watching TV broadcasts and either only watch DVDs or stop watching TV altogether. Keep in mind that many TV shows are now available on DVD approximately four or five months after the last season episode airs.

One final alternative is to subscribe to Netflix (www.netflix.com) or a similar DVD rental program, through which you receive between one and three DVD rentals at a time (movies, documentaries, and TV shows), sent to your home. Returns are free, and as soon as you return a DVD, another is sent to you right away, based on a list you create of hundreds or thousands of DVDs you want to rent, in order of priority. Like Hulu, you can also stream movies and TV shows either to your computer or your television set. Plans run from about $6 per month to $20 per month.

Alternatives to a Landline Phone

The obvious alternative to a landline is a cell phone. However, if getting rid of your landline will increase your need for cell minutes (and you will pay a higher fee for that), double check your math. Find out what other cell companies are offering their customers: free incoming calls; free nights and weekends; nighttime rates starting earlier than 9:00 P.M.; free calls to customers on the same network; free text messaging; and so on. When your contract is up, if you can switch to another cell company and save money by not having a landline, go for it!

Avoiding Extended Warranties

No matter what major purchase you make—car, furnace, computer, or dishwasher—you'll probably be offered an extended warranty by the company selling you the product. For "just" $79, you can add an extra year to the existing warranty. Sometimes you can even add three or four years of protection. These extended warranties can be a good investment in some cases, but they're a bad idea at other times.

When Not to Buy an Extended Warranty

If any of the following apply, steer clear of an extended warranty:

- You intend to own the product for only as long as the original warranty is in effect.
- Within a few years, the product will be out of date, and you'll want or need to get a better, more powerful model.
- The purchase price is low enough that you wouldn't be strapped if you had to buy another in a few years. Repairing this product is simple and inexpensive.
- The cost of this product is likely to decrease over the next few years. The extended warranty costs more than 20 percent of the purchase price.

When to Buy an Extended Warranty

Do get an extended warranty if any of the following is true for you:

- This piece of equipment is critical to your livelihood.
- You know you can't afford to replace the product if it breaks.
- The warranty is a very good deal.

You can set up your own "extended warranty" savings account. If the product you're buying costs $120 and comes with a one-year warranty, put $10 per month into your savings account. When the year is up, you'll have enough money in savings to cover the purchase of a new product, should the old one break. If the cost of the product tends to go up with time, as is the case with a car, put a little more than the existing cost into your savings account.

Buy Reliable, High-Quality Products

This idea may seem to go against most money-saving advice, but the truth is that high-quality products tend to last longer. If you buy a well-researched, reliable car instead of an inexpensive economy car, you'll pay substantially more. But if the economy car fizzles in three years and the Volvo keeps running for fifteen years after that, you'll probably save money in the long run.

Keep the following tips in mind, however, when shopping for quality:

- **If buying the quality item will wreck your budget** either save up and come back when you can afford it or make do with the less expensive item.
- **Don't automatically assume that higher price equals higher quality.** Sometimes higher prices are simply the result of savvy businesspeople thinking that consumers will associate their products with quality if they charge a lot.
- **If you aren't sure how to recognize reliability and quality**, check out Consumer Reports, your best source for honest, detailed testing results for tens of thousands of products. Go to their website at www.consumerreports.org. If Consumer Reports thinks a product has problems with quality, keep shopping. Because they don't accept advertising dollars; their testing results are unbiased. Most libraries have subscriptions to this publication, so if you're willing to do a bit of research, you can get the information for free. Note also that many companies

include space on their websites for consumer reviews of products. Don't be afraid to read as many of these as possible before making a purchase.

■ **Don't worry about buying a quality product** if you're not planning to keep it very long. If you're on vacation and forget your swimsuit, don't spend a lot for another one—just buy something that will see you through.

Don't Go Trendy

Before you buy anything, ask yourself whether you're buying it because it's the best-quality item you can get for the price or because it's a hip, happening item that makes you feel good for the moment. Women's shoes and purses come to mind as short-term, trendy items that tend to be out of style in a year or two.

Many budgets are blown on novelty items, and what's so frustrating about buying them is that a couple of weeks or months later, you can't figure out what you saw in the item in the first place! Before you buy anything, apply the one-year test: Is this an item you'll want a year from now? If not, pass it up.

Buy Secondhand

Secondhand doesn't have to mean "inferior." Although you should exercise some caution when purchasing secondhand items, you can often find bargains. There's no reason to spend $25,000 on a new car if you can get a good secondhand one with relatively low mileage for a quarter of the price. Amazon will let you purchase secondhand books and will even evaluate their condition for you. You can find secondhand clothing in thrift stores. And eBay can be a treasure trove of bargains, provided that you know what you're looking for and what matters to you about it.

Becoming a Late Adopter

You don't have to be the first kid on your block to get everything. Personal electronics, especially, tend to have a high initial price, and then settle into a lower price for late adopters. For example, iPods of all shapes and sizes are now available at Sam's Club or Costco for less than you'd pay elsewhere, but it takes a while for the latest models to reach the discount stores. Give it a few months, and then make your purchase—if you budgeted for the item!

Shopping Tag Sales, Resale Shops, and Online Auctions

Whether you're furnishing a nursery or building a wardrobe, tag sales (also called garage sales or yard sales) and resale shops—including those from Goodwill Industries and The Salvation Army—can save you a bundle.

Does this go against the advice in the preceding section to buy high-quality items? Not necessarily. Just because an item is being sold at a tag sale or resale shop doesn't mean it isn't a high-quality item. The mere fact that the item has lasted long enough to be worn by someone until it no longer fit, went out of fashion, or the person became bored with it points to the fact that this is a long-lasting product. Cheaply made products don't usually end up at tag sales and resale shops—instead, they get thrown out.

Some low-quality items do appear, however, so you need to know a couple of tricks for shopping at tag sales and resale shops. These techniques are discussed in the following sections.

Preparing a Shopping List Ahead of Time

If you just go to browse, you're likely to end up buying something that you don't need; even the deepest discount isn't a bargain if you don't need the item. Before you leave home, determine your needs and put them down on paper—and then don't buy anything that's not on your list, no matter how wonderful or how cheap it is.

Purchasing High-Quality, Undamaged Products

If you're interested in an item, pick it up and carry it with you. If you're not sure you want it and don't pick it up, it's liable to be gone when you go back to look for it, especially at a tag sale. At large resale shops like Goodwill Industries and The Salvation Army, you may never again find that blue shirt among the hundreds of blue shirts they stock.

After you've looked at everything you're interested in, turn to the items you've been carrying. Look closely at any product before buying it. Examine it for damage of any sort; turn it over and inside out to see whether it's cheaply made or is something that will last a while. Keep in mind that even buying a $2 chair or a $1 pair of pants isn't a good deal if it breaks or rips the first time you use it.

Consider Haggling—or Not

A lot of people haggle at garage sales. Most people holding the sales expect it, but the choice is up to you. You may save a few bucks, but the person having the garage sale is also trying to make some money, so if the marked price seems acceptable to you, pay it.

Winning at Auctions

If there's an item you've been looking for but can't quite afford, get yourself over to an online auction site (such as eBay, at www.ebay.com) to see whether anyone is offering it at less-than-retail value. You can search for items by keywords (better than browsing, which is too tempting), and you may be given two options: an option to submit a bid, and an option to buy it immediately at a set price. If that buy-it-now price is lower than what you'd pay elsewhere, be sure to check what the shipping and handling charges will be.

If you decide to bid on an item, be sure to utilize the automatic bidding function that will keep automatically raising your bid until you reach your highest price. This will keep you from having to be notified each time someone outbids you; it also forces you to set your highest price well in advance, so that you don't get caught up in a bidding euphoria and blow your budget. If you're afraid you'll be tempted to bid higher than the limit you originally set, be sure to be away from email in the final minutes before the auction expires.

Reducing Gift Expenditures

Contrary to popular belief, you don't have to purchase gifts for your friends, family, and coworkers on every birthday, anniversary, or Hallmark holiday. You can, for example, save money on holiday gifts by drawing names among your friends, family, and/or coworkers. For any occasion, you can give a small donation to a favorite charity in the name of the gift recipient and send a card explaining the gift. Consider making gifts as well: cookies, breads, soups, and so on. Also, anyone—from a close friend, to a casual acquaintance, to a family member you don't see often—will appreciate a simple handwritten note from you.

One way to save on gifts is not to give them at all! Let friends, family, and coworkers know that while you're getting your finances under control, you won't be giving—and don't expect to get—presents for the next few years. Send a note to this effect in late September or early October to give people time to adjust!

Avoiding Dry-Cleaning

Unless the product you're cleaning absolutely, positively has to be dry-cleaned (check the label), don't use this expensive service. Many articles of clothing—even silk, wool, and linen—can be hand washed or washed using the delicate cycle of your washer using an extra-mild detergent.

Opportunities for saving money around the household constantly turn up. Keep a list on the refrigerator and add to it as you think of other ways to cut down on household expenses.

8

SAVING ON TRANSPORTATION

With the price of gasoline fluctuating over the past several years, many people have found that transportation has become a more and more significant element in their personal budgets. Long gone are the days of cheap oil when we could practically fill our tanks with the spare change rattling around in our pockets. Now, as you draw up and re-evaluate your budget and look for savings, transportation is a useful place to go.

How Many Cars Do You Need?

Many middle-class American families developed the idea (mainly in the 1990s) that for every person in the household of driving age, there should be one car. This led many families to own three or even four vehicles. As of 2009, more than a third of U.S. households owned two cars, and almost 20 percent owned three or more. That's a lot of cars—and a lot of expense.

If you and your family own two or more cars, begin by asking yourself, how many do you really need? Sure it might be a bit inconvenient for you to have to drive your spouse to appointments or to discover that one of the kids has the car for a date just when you need it to go to the store. But with some cooperation and effort, it should be possible to work out a single-car schedule that meets everyone's needs and saves you money in gas, insurance, and maintenance. Then there's the question of whether you actually need a car at all. After all, cars can cost a lot of money: Payments or leases usually run several hundred dollars a month; maintenance and repairs are expensive; over-the-top gasoline prices can squeeze your budget; and registration and insurance can set you back a thousand dollars or more each year.

Looking Into Public Transportation

If you live in an area where you can walk or bike to work and the grocery store, or if you have a reliable mass-transit system in your area, consider getting rid of your car. To most people, this is a revolutionary—if not repulsive—idea. Having a car is like having a name: Everybody has one! Well, actually, they don't. Plenty of people who live in large cities don't own cars, and they love it. And more and more environmentalists are touting the benefits of walking or biking or riding public transportation to work, so you're not completely alone there, either.

Budgeting Tip: Many people wonder how you'll get home for the holidays or take vacations if you don't have a car. The simplest solution is to rent a car when you need one. You may pay a lot for the rental five or six times a year, but that cost won't come close to the amount you now pay in car payments, insurance, maintenance, and so on.

Even if you're not a city dweller or staunch friend of the earth, getting rid of your car can make sense. There's an immediate financial impact. If you're making monthly car payments, those will stop right away. And if your car is paid off, you'll get some cash that will help you pay your other bills.

Carpooling

Many cities strongly encourage carpooling through the construction of high-occupancy vehicle (HOV) lanes. These lanes are open only to vehicles carrying two or more people and often speed along quickly, bypassing single-driver vehicles sitting and sweating in heavy traffic. Carpooling is most effective when you can do it with people who travel regularly to the same place—mostly those with whom you work. You can send around an email or stick a sign up on the bulletin board near the water cooler. Find out who from your workplace lives near you and would be interested in saving some money on transportation. Then decide upon a fair amount everyone will kick in for gas.

Of course, you've got to agree that you'll all be ready to go at the same time and that you'll all leave work together. But with a little effort, carpooling can save you quite a bit of money in transportation costs. Some employers encourage carpooling by their employees, offering vouchers and other incentives for those who carpool.

Going the Bicycle or Vespa Route

Even if public transportation in your area isn't up to par, you may still be able to live without a car, especially if you live in an area with a mild climate. Cycling to work every day gives you two immediate benefits: 1) It saves you money; and 2) it gets you into shape. Many companies now offer a shower at work, so if you get sweaty on the ride in, you can shower and change into work clothes when you get there. By installing a pack on your bike (called townies) that holds two sacks of groceries, you can also stop by the store on your way home.

If being completely reliant on your physical prowess to get you around town is a little much for you, consider investing in a moped, such as the popular Vespas. These economical vehicles are like low-powered motorcycles and generally run from $2,000 (used) to about $3,500 (gleaming and new). If you can sell your car, buy a Vespa, and save bundles on gas, insurance, and registration, it might make getting a little windblown not seem so bad.

Keeping a Paid-Off, Reliable Car

Note that if you have a reliable car that's paid off, runs well, and costs a reasonable amount in gasoline, maintenance, and insurance, you're probably better off not selling it. A car like this is just too rare to part with.

Budgeting Tip: Next time you buy a car, purchase the highest quality model you can afford, put as much money down on it as you can, and arrange for the fewest number of payments possible. Then plan to drive the car—payment free—for as many years as you can after you pay it off.

Another situation in which selling your car isn't a good idea is if you're upside down in your loan—that means your car is worth less than you owe on it. If you're upside down in your loan and interest rates are lower than they were when you bought the car, look into refinancing your car loan.

Deciding Between a Hummer and a MINI

If you're in the market for a car (new or used), you'll need to make an important decision: Hummer or MINI. Another way to put this is: are you going to go with a roomy interior and low mpg, or a small interior and high mpg? The MINI Cooper (and similar small or hybrid cars) offers a few advantages:

- **They cost less.** Sure, you can get a souped-up MINI (the S version, in the convertible, with all available packages and options), but even that's not going to cost you as much as a Hummer. If you finance your car, this means that your monthly payments will be lower or you'll be able to finance your car for fewer months.

- **They get much better mileage.** This has substantial financial ramifications over the next several years, especially given the recent jack in gas prices.

- **Your auto insurance may be cheaper.** This didn't used to be the case, as smaller cars were also often less safe, so insurance for smaller cars wasn't any less than for larger ones. But today's small cars often do just as well in crash tests as larger, more expensive cars, and because the smaller cars cost less to replace, insurance companies charge less in premiums.

- You can park in all those parking garage spaces that say "Compact cars only."

Choosing Between Leasing and Buying

Except for a few business-related tax breaks, leasing a car will never improve your financial picture. Leasing a car amounts to borrowing it for a specified number of months or years and then, at the end of your contract, giving it back. Leasing is attractive to many people because your monthly payments are lower than when you buy and the length of a lease contract is usually fairly short, which means you can get a new car more often than if you buy. But leasing it is really just having a long-term rental car.

Budgeting Tip: If you must own a car, don't lease! Instead, buy a reliable car on the fewest number of payments you can afford and plan to drive it for ten years—or more. After you've paid it off, keep making the payments to your savings account so that you can pay cash for your next car.

In order to improve your financial picture, stop thinking of a car as an extension of who you are. Ultimately, if you're miserable because you're sinking deeper and deeper into debt and don't know how you're going to pay your bills this month, who cares what you're driving? You also want to stop thinking of a car payment as a fact of life. Just imagine how much more breathing room you'd have each month if you didn't have a car payment. Well, leasing never lets you go there. You're locked in to making a payment every month, and when you're done paying, you still don't own a car. You just have to go out and get another one, and make the lease payments on it for several more years.

Shop Around for Car Insurance

Consider the following story: A Midwestern couple who lives in a small town pays nearly $1,150 per year for insurance on their two cars. They don't drive far to work, haven't had any accidents or received any speeding tickets, and they own their own home, so they're a good risk. They think $1,150 per year isn't too much to spend—it's on par with what their neighbors pay, and it's less than they paid when they lived in a larger city. They have an agent, but they haven't seen or spoken to her in years, and if they do have an accident, they are supposed to call a toll-free number, not the agent.

Then they see an ad for another insurance company. They call the number to get its price and the price of three competing insurance companies. (Here's where the story gets a little unbelievable.) For the exact same coverage, that company offers them insurance for $385 per year, a savings of $765 per year! You're thinking this is a made-up story, right? Nope—all true. This scenario happened about five years ago, and the service and coverage has been exactly the same—maybe even better because the company they went with has a twenty-four-hour customer service hotline that's operated seven days a week. The moral of the story is this: Shop around. Insurance companies change their products and prices all the time. Once a year, do a quick internet search of insurance prices, and if you find a substantially lower price, ask your current agent to requote your policy to see whether he or she can match what you've found.

Keep in mind that speeding, reckless driving, and driving under the influence can ruin your finances. Not only will you get socked with the soaring cost of tickets, but your insurance rates could double or triple. So drive carefully and safely. If you maintain a good driving record, you may be eligible for a Good Driver Discount at your insurance company. Ask your agent about it.

Comparison Shopping

You have nothing to lose—and potentially a lot of money to gain—by contacting Progressive Insurance at 800-PROGRESSIVE or www.progressive.com. Keep in mind that if you've received a lot of traffic tickets, have been in one or more accidents, live in an area that tends to produce a lot of accidents or car thefts, drive a car that's expensive to repair, or have a very long commute, your insurance payment may be quite a bit higher than the example given here. But because Progressive gives you the insurance rate for its major competitors, you may find a non-Progressive rate that's still better than what you're paying now. Use Worksheet 8-1 as a handy place to compare rates and coverage.

Worksheet 8-1: Insurance Comparisons

Insurance Company	Semiannual Premium	Type of Coverage
	$	
	$	
	$	
	$	
	$	
	$	
	$	
	$	
	$	
	$	
	$	
	$	

Raising Your Deductibles

If your insurance payments are still uncomfortably high after you shop around, try raising your deductibles (the amount you pay out of pocket if you have an accident, your car is stolen, or a flood washes your car away). You can save quite a bit on your annual insurance costs by increasing your deductibles from $250 to $500 or from $500 to $1,000 (per incident). Some companies don't offer high deductibles, but if yours does, see how much of a difference raising it can make. Do be sure, though, to put the amount of your deductible in a savings account so the money is there if you need to repair your car.

OTHER TIPS ON SAVING

Pick Nature Over Electronics

Instead of spending endless money on electronics, which often sucks your money (let's face it—Microsoft and Sony are always getting you with those add-on expansion packs or super cool new gadget/game), put your wallet way and go enjoy something free—nature! It's better than video games or being online because you can enjoy fresh air, watch animals, and exercise in a fun way. It's also family friendly and you can enjoy nature almost anywhere! Best of all, it's free!

Explore Your Own Backyard

Instead of planning lavish vacations in tourist spots, which are almost guaranteed to suck your money away, plan a "staycation" and check out the local scene. Oftentimes, there are amazing, cheap things to do in your own town that you never knew existed. See if your area has any great museums, historical sites, parks, natural trails, etc. Having a staycation means that you will spend less on travel and hotel fare and, if you're lucky, you'll discover some cheap and fun things to do right in your own backyard.

Avoid Malls Without Money

Don't even think about going to the mall unless you have extra spending money. How many times have you tagged along with a friend "just to look" or "just to go along" and ended up buying something that you saw and wanted? Oftentimes, that happens to people who don't plan on spending money, so it usually comes right out of their pay for the month. Shopping anywhere without money is usually dangerous for your wallet, so only try to go when you actually have an allotted amount of "recreational shopping" funds or simply save it for a special reward or treat later on.

Record Shows and Skip Through Commercials to Avoid Spending Urges

When watching commercials, your chances of being tempted to buy something can greatly increase, thanks to marketers. To avoid risking the urge of spending from commercials, record your favorite TV shows in advance and watch them later. This way, you can fast forward through the commercials. I almost never watch TV live anymore! When I was younger I would be forced to sit through who knows how many hours of different commercials. Modern day technology truly is incredible when used wisely!

Buy Safe Used Cars From Individuals

Instead of buying a used car from a dealership, try looking around to see if you can buy one from an individual seller. Car dealerships often try to sell you those "extras" such as extended warranties or cleaning services which will rack up your total bill. If you choose to finance a car through a dealership, you will still get hit with a couple hundred dollars of interest in the long run. If possible, shop around before heading to a dealership. You can usually find good deals from individual sellers in your local newspaper classifieds or on Craigslist. This, however, is a risky proposition. If you're lucky, you'll save thousands of dollars in the long run, and if you're unlucky, it could cost you thousands of dollars. My actual personal preference is to buy new cars that are known for quality and to just take really good care of them so that they last a very long time.

Take Care of Personal Possessions (toys, cars, lawn, etc.)

By taking care of your personal possessions, you can make them last longer, therefore getting more value out of your money. Take care of everything that you have. Take care of your house so that it doesn't end up in need of a total makeover. Teach your children to take care of their toys so you won't be forced to buy replacements. Take care of your car so that you can have many years to come with no car payment for repairs.

Re-purpose Items

Many of your personal possessions can be re-purposed into something great and useful just when you think you no longer have a need for it. For example, you might have a special t-shirt that you wore as a child and eventually you grew out of it. One cool idea is to turn it into a pillow that you can use to decorate your house or sleep on. I've also seen a cool idea to turn your child's crib into a desk once they've outgrown it. The possibilities are endless!

Watch Bank Statements

A great way to better manage and save money is to keep an eye on your bank statements, especially if you let somebody else (kids, spouse, etc.) use your credit and debit cards from time to time. When you use plastic and auto-debits to pay for things rather than cash, it is very easy to lose track of just how much you're spending. At the end of the month, review them. This way, you can have a good idea of how much money you're spending each month and you can see if anyone else is racking up your bills.

Cell Phone Plans

Have you reviewed your cell phone plan lately? Big-name cell phone companies like Verizon and AT&T charge high amounts for cell phone contracts that you have to lock yourself. However, there are other options that you can consider for your cell phone plan, such as Boost Mobile. Their company doesn't lock you into a contract and you can activate almost any phone for a reasonable price. Basic cell phone payments can go down to as low as $30 after you make a certain number of payments on time and smart phone payments can go as low to $45-50. If your cell phone bill is killing you, that might be something to look into.

Actively Ask For A Discount or Deal

You may be shocked at how much you can save if you simply ask for it. I go out of my way once a year to make sure I'm getting the best deal possible on everything that I am spending money on. Some areas to focus on would be your cell phone bills, cable bills, credit cards and car insurance. Especially with the very big companies, you can usually find a customer service representative that will give you a discount on your bill. If the first person says no, try again a later time. Also, don't be afraid to ask for the manager. I harassed my cell phone provider quite a few times before I finally found a nice representative who gave me an incredible deal! I also called my cable company around five times before I got a deal so good that many say it's the best they've ever seen!

Be persistent and be sure to shop around. Many companies will match discounts from other companies or give you an incentive to stay if you threaten to cancel or leave.

9

WATCHING
WHAT YOU SPEND

I f your spending is getting the best of you and creating more and more debt for your family, try freezing your spending for the next several months. Freezing your spending isn't easy, but it can stop your accelerating debt dead in its tracks.

What Freezing Really Means

Freezing means going cold turkey on your spending—you temporarily stop buying. For the short term, you cut out all but the most essential spending; your cuts will include personal appliances, home appliances, clothing, shoes, CDs, DVDs, decorative items, linens, computer accessories, and so on. You freeze your spending for a predetermined amount of time—usually six to twelve months—and just stop shopping. Of course, you can still buy groceries and the required supplies for your home, but you don't buy anything else.

Some people believe that they must spend in order to keep the American economy going. While consumer spending does impact how much money many businesses make, your six or nine months of thriftiness is not going to spin the economy into an uncontrolled recession. Besides, you'll be back someday.

Reducing Temptation During a Freeze

People who temporarily freeze their spending usually find that the best way to stay the course is to steer clear of opportunities to spend money:

- Don't read the ads that come in the Sunday paper.
- Don't stop at outlet malls when you travel.
- Dispose of all the catalogs you have in your possession.
- Call all the companies that send you catalogs and have them both remove your name from their mailing lists and stop selling your name to other companies.
- Don't visit internet sites that sell products.
- Don't go to the shopping mall food court for a quick meal.
- Don't meet friends for an afternoon at the mall or any other store.
- Don't go window shopping at an appliance, music, or computer store.
- Discontinue any music or book clubs, even if you have to buy your remaining required purchases to do so.
- When grocery shopping, don't inadvertently wander into the consumer goods section of the store.
- Send gift certificates instead of actual purchases as gifts so that you don't have to go to a store or browse a catalog or website.

The following sections help you freeze your spending a little less painfully.

Establishing What's Really a Need

Understanding the difference between a need and a want is really the crux of sorting out your financial difficulties. In an effort to make ourselves feel better about being consumers, we continually elevate wants to the level of needs. But we actually have few needs, at least in the realm of products that you can buy:

- Shelter
- Clothing
- Food and water

Thousands of years ago, this list meant a mud, straw, or wooden hut, along with some animal skins and just enough calories to survive. Today, we have escalated these basic human needs, and they have become so intertwined with wants that we're not sure how to separate them. Yes, you need shelter, but you do not need a four-bedroom home with a formal dining room, a fireplace in the great room, a three-car garage, a kitchen with cherry cabinets, and a bonus room over the garage. That's a want.

The same is true for clothing. Humans need a way to stay warm and dry, but they do not need ten suits or eight pairs of jeans. Those are wants. And while everyone needs food and water to survive, that food does not have to come from a five-star restaurant. You also only need enough calories to survive, not enough to add three to five pounds each year, as the average American does.

The desire to own and consume is very strong in Americans, and it enables us to justify nearly any purchase in the name of needs. Don't buy into it. Instead, use Worksheet 9-1 to list every need you have (you might want to use a pencil, though, and keep a good eraser handy). Be very specific in your list: Don't just list "house"; instead, write a description of the house you need and the amount it will cost.

Worksheet 9-1: Needs versus Wants

Need (Description)	Cost	Consequences of Not Buying
	$	
	$	
	$	
	$	
	$	
	$	
	$	
	$	
	$	

After identifying the consequences of not meeting a need after you've listed all your needs, identify what would happen to you if you didn't get each one, asking yourself the following questions:

- Would you or others around you die?
- Would you or others suffer physical pain or extreme physical discomfort?
- Would your health or the health of others suffer in the long term?
- Do you know for sure that you would lose your job without this item?

If none of these would happen, it isn't a need, it's a want, and you have no business buying it during a spending freeze. Remember this the next time your mind tries to talk your wallet into giving in.

Establish—and stick to—a shopping list for your needs before you leave the house and head out to spend money (which is likely to include only groceries and toiletries). Be sure that they're needs, and don't pad the list because you're in the mood to buy. Keep in mind that you are probably feeling deprived, so you may try to satisfy your spending itch by splurging on groceries and toiletries.

Before you leave for the store, write down everything you need to get, and also scribble in an estimate of how much each item will cost. Then total the bill. If it's less than you planned to spend, stop writing out your list and immediately go to the store. If the total is more than you planned to spend, begin crossing items off your list before you go until you get down to the budgeted amount.

Budgeting Tip: Don't justify veering from the list because something is "such a good deal." Instead, remember that the best possible deal is to spend $0, so even if an item is half price, you can't buy it unless it's on your list.

Then, buy only the items on the list. Don't add items to the list and then cross them off while you're standing in the checkout lane. Instead, stick absolutely to your list. If you see something you're sure you need but it isn't on your list, put it on next week's list when you get home. Today, you can buy only what's on your list. Be vigilant about this process, and you'll never overspend on groceries and toiletries again.

Put Away Your Credit Cards

No, seriously, put them away for at least six months. Put them in a safe place that's hard to get to, such as a safe deposit box at the bank (which will probably cost around $20 per year, an amount that's worth spending if it keeps you from getting further into debt). The farther away the credit cards are from you, the better.

For six months, pay for all of your day-to-day purchases with cash and pay your bills with a check. When you're shopping for purchases that are allowed—such as groceries and toiletries—write out a list before you go, estimate how much you'll need, and take no more than $10 over that amount. When you're not supposed to be making any purchases, limit the amount of cash you carry around to $5 and a few quarters. That will allow you to pay for parking if you need to, but not lunch or a flat screen TV!

Tuck Away Your Debit Card

Although a debit card is technically like cash or a check, in reality it feels much more like a credit card. Because you don't hand over cash, you may feel as though you're not really paying for this purchase, much like when you use a credit card. And if those funds are earmarked for other purposes (like paying off your debt or saving for a vacation), you'll end up without enough money set aside for those items by the end of the month.

If you take $80 in cash to the grocery store, you'll be very careful not to exceed that amount with convenience foods. But if you take a debit card, you're not likely to be nearly as careful. Put the debit card in the same place you put the credit cards—your best bet is in a safe deposit box.

Create a Wish List

A wish list is an outlet for your hot little fingers and creative mind while you're in a spending freeze. The basic idea is that you write down everything you'd ever like to buy. The list may range from a new

TV to whitening strips for your teeth to a sailboat. Anything you're not allowed to buy during a spending freeze is fair game. Nothing on the list has to be sensible or practical or a wise financial decision.

Sometimes when you're not spending, you feel disconnected from our consumer-oriented society, and a wish list makes you feel like your old self again. When you feel the itch to spend, go online or look at a friend's catalogs and write down the item number, description, page number, and so on, of any item that looks interesting. Act as if you're really going to buy the item. But don't. Just add the item to your list and let the list sit for a while. The act of writing the item down will feel, strangely enough, very similar to how you feel when you actually buy something. It sounds completely crazy, but it works!

When you brainstorm your wish list, think "pie in the sky." You're just daydreaming right now—later, you can make your list more realistic. So write down whatever you can imagine in your future. But make sure it's your wish list. Don't put a sailboat on your list if you really don't like water!

Pare Down the Wish List

Just listing the items can be cathartic when you want to buy, buy, buy. But listing the items on Worksheet 9-2 can also help you cross some items off the list. When you write down an item's name and cost, also check off one of the three needs categories: "Need Today," "Need This Month," or "Would Like Someday." If none applies, don't check anything off. Tomorrow, revisit any item that you indicated you needed today. Is the need still strong? In a month, review any items that you needed this month, and also look at the items that you'd like someday. Do you still feel strongly about them? Cross off any item you no longer feel you need and check off new categories for some items.

Worksheet 9-2: Your Wish List

Item Name	Cost	Need Today?	Need This Month?	Would Like Someday?
	$			
	$			
	$			
	$			
	$			
	$			
	$			
	$			
	$			
	$			
	$			
	$			
	$			
	$			
	$			
	$			
	$			
	$			
	$			
	$			
	$			

Reviewing a Sample Wish List

Your wish list may look like Table 9-3:

Table 9-3: Sample Wish List

Item Name	Cost	Need Today?	Need This Month?	Would Like Someday?

Now, suppose thirty days have gone by, and the list looks like Table 9-4:

Table 9-4: Sample Wish List, Round Two

Item Name	Cost	Need Today?	Need This Month?	Would Like Someday?

Thirty days later, the list may look like Table 9-5:

Table 9-5: Sample Wish List, Round Three

Item Name	Cost	Need Today?	Need This Month?	Would Like Someday?

At this point, you've narrowed your list to items you would clearly like to own and can begin to save for when your spending freeze is over. You also have a ready-made list if anyone asks you what you really want for your birthday.

10

PAYING DOWN DEBT

One of the important parts of responsible budgeting is getting a good grip on what you owe. This can be disconcerting, especially if you've not paid attention to the amount of your debt for a while. Some families are shocked by the amount of debt they're carrying and panic when it's all written down in one place. This is understandable—but never fear. We can find a way out of this morass of debt.

If your debts are crushing you, you may need to take action to restructure them through a number of means: credit counseling; debt consolidation; or selling some of your assets.

Budgeting Tip: The amount of debt you can comfortably carry depends on many things, chief among which is your income. If you can comfortably make all your debt payments and are gradually reducing the amount of debt you carry, you're in pretty good shape. On the other hand, if the amount of debt you have is either increasing or staying steady, you need to take steps to reduce it.

Understand How Debt Is Restructured

Does this sound like you? You have too much debt to handle—maybe you've charged more than you can afford on several credit cards, you have school loans plus a car and house payment, and the usual payments for utilities and food. You're having trouble making monthly payments, perhaps you are already a few months behind, and you're starting to be (or have been for some time) hassled by debt collectors. If so, debt restructuring is exactly what you need! The idea is that you change the way your debt is structured by lowering interest rates, lengthening repayment schedules, combining several payments into one smaller payment, or getting some of the debts forgiven, and, at the same time, you stop getting further into debt. You may have to give something up, but you'll probably come out way ahead in the long run.

Budgeting Tip: *If you've been using a check-cashing service to get cash for your paycheck (or a cash loan against your next paycheck), stop immediately! Most of these companies charge a ridiculous amount of money for their services. Instead, open a bank account (look for a totally free one), which you can open with anywhere from $5 to $50. Never take a loan from a check-cashing service, as you'll find yourself paying an exorbitant rate of interest.*

There are a number of ways to restructure your current debts. You might see a credit counselor to discuss your options (this is often a good place to begin because it's usually free), consolidate most of your debts into one payment, sell some of your assets, use the equity in your house to pay off your debts, or declare bankruptcy.

Get Credit Counseling

Credit counseling is usually a free, nonprofit service that offers an alternative to bankruptcy. Each agency assigns you a counselor who reviews your debts, assets, income, and so on, to help you identify your options other than bankruptcy. Sometimes the credit counselors are bona fide financial gurus, but more often they're simply well-trained, well-meaning volunteers who offer an excellent service. All credit-counseling agencies offer their services in complete confidentiality and may offer services over the phone and internet, as well as in face-to-face consultations.

Keep in mind, however, that not all credit-counseling agencies are nonprofit, and some are almost like scams. You can find out more in the "Consolidating Your Debt" section in this chapter.

Make Sure the Counseling Is Free

Your initial counseling session(s) should be completely free. If it isn't, get out as fast as you can! Many wonderful nonprofit credit counseling agencies exist, so don't waste your money on an agency that charges you for counseling. While you may have to pay a small fee to consolidate your debt, the counseling session itself—in which your finances are sorted out and advice is offered—should be free.

Get Comfortable with Your Counselor

Be sure you trust your counselor and feel confident in his or her abilities. If you don't, find out whether you can have another counselor assigned to you. Keep in mind, however, that your agency is probably a nonprofit organization with limited resources. You should have a darned good reason for wanting to be assigned a new counselor before you ask for this special treatment.

Take Advantage of Free Financial Education Opportunities

Credit counseling agencies often offer free short seminars or informational brochures on how to get out of debt, manage money, save for a down payment on a house, save for retirement, and so on. They do this as a service to the community, like any nonprofit agency may do.

If you're not ready to speak to a counselor but want more information, consider attending one of these seminars. There you'll meet one or more of the counselors who work for the agency, and you may become more comfortable with the idea of confiding in this perfect stranger. Credit counselors are listed on numerous sites on the internet.

Consolidating Your Debt

The biggest challenge with debt is not the simple amount you owe (called the principal); it's the interest on that amount. The interest is what keeps you paying, often for years, until you've paid far more than you originally borrowed.

Instead of writing a separate check for the minimum amount to all of your creditors, all of that unsecured debt (debt that doesn't have a sellable item, like a house or a car, attached to it) can be turned into one payment—usually at a much lower interest rate—that you can more easily manage each month. This is what debt consolidation is all about.

Debt consolidation is not a loan, nor is it a forgiveness of your debts: You do pay off all your debts in due time. But debt consolidation often offers a lower interest rate than you're currently being charged, and if your debts are with collection agencies who expect immediate payment, you may be able to take more time to pay those debts. The best part? The harassing phone calls and letters will stop immediately.

Usually, you sign an agreement in which you allow your credit counselor to contact your creditors, let your counselor submit a budget on your behalf, agree to make your new payment on time (or have your payments automatically withdrawn from your checking or savings account), and agree not to get into further debt. If your creditors agree (and they usually do), you are usually in a position to be free of these debts in three to six years, provided your budget allows for this. Your credit counselor will put you on a tight budget until your debt is paid off.

Using an Accredited Agency

Most, but not all, debt consolidation is performed by credit-counseling agencies. Before you sign on with any agency, check with the Council on Accreditation of Services for Families and Children, the National Foundation for Credit Counseling, and the Better Business Bureau.

Remember that not all credit counselors are the same! Credit card companies have always appreciated credit counseling services that help people figure out how to pay back their debts, even if it takes them a long time. This is because when cardholders file for the alternative—bankruptcy—the credit card companies often get no payment at all. So, to help these nonprofits, credit card companies sometimes donate a percentage of the card balance to the credit-counseling service. Some entrepreneurs, hungry for the fee that credit card companies give for credit counseling, have started for-profit businesses that advertise as credit counselors. But these companies often push the consumer to pay the credit card companies first—or worse, will work only with debts owed to creditors that pay them—which may not be in your best interest.

Examine the Fee—If Any—for Debt Consolidation

Most nonprofits charge a nominal fee for debt consolidation. Expect this, but do not pay more than $25 per month for this service. The money goes to a good cause—paying the agency's considerable expenses—and because of your lower interest rates, you'll still save a bundle of money.

Review the Terms of Your Agreement

Be sure you read the terms of your agreement carefully. You'll usually be expected to make your monthly payment on time—with no exceptions—and you'll also agree not to get into any more debt. This is a bit of tough love because, ultimately, you can't break your cycle of debt if your credit counseling agency bails you out and then you get right back into debt again. Because they're going to all the trouble of intervening on your behalf, you have to agree to change your lifestyle. It's a tall order, but it's the only way most credit counseling agencies will work.

Consolidate Your Debts on Your Own

Another way to consolidate your debts is to use one of your credit cards to pay off all your other debts. Many credit cards even provide checks or special forms that help simplify this process.

Under most circumstances, however, consolidating this way—on your own and without the guidance and support of a counselor—isn't a good idea. Because you won't have signed an agreement not to rack up any more debt, you may be tempted to use your now-paid-off credit cards to spend more money, making your situation worse.

In addition, even if one of your credit cards is offering a low interest rate to transfer the balances from your other cards to theirs, the rate is usually good only for a limited amount of time (like six months) and may skyrocket after that. Many credit cards also charge a transaction fee to pay off the balances of other cards. A credit counselor can usually arrange for an even lower interest rate for your debts—and it won't expire.

Sell Some Assets

Besides debt consolidation, there is another way to raise money to pay off your debts: Sell some of your assets. If your house, apartment, storage unit, or parent's house is stocked with items belonging to you that you no longer use and that may have some resale value, consider selling them and using the money to pay down your debts. (This is where you see the value of the list of your assets we made at the beginning of this book. It gives you a clear idea of exactly what you own that you might be able to convert into cash in order to reduce your debt.)

Budgeting Tip: Don't confuse pawnshops with tag sales, where you drag out all your stuff and try to make a few bucks. Let's be clear: The people who run pawnshops are nearly always loan sharks, often charging as much as a 25 percent annual interest rate. Steer clear!

Selling Valuable Items

Items that you may be able to sell—and that may be valuable—include furniture, jewelry, an automobile or motorcycle, exercise equipment, recreational toys (pool table, bike), paintings, signed books, computer equipment, guns, memorabilia (baseball cards, signed sports balls), coin or stamp collections, and outdoor equipment (grill, riding mower).

Whatever items you plan to sell, if you expect a high price for them, make sure they're in excellent condition. If they're not nearly new, consider holding a tag sale (see the following section).

Don't sell any items that don't belong to you! That may be considered theft, landing you in hot water. Also, don't sell anything that has a loan against it unless you plan to pay off the loan that same day. Contact your lender about how to sell an item that they hold a lien against.

You can sell valuable items in a variety of ways:

- **Advertise in your local newspaper or a pay-only-if-you-sell publication.** Although an ad in your local paper can be a bit pricey, you won't have to mess with shipping the item to an out-of-town buyer. Classified ads in some papers are quite inexpensive—and most feature a searchable internet component. The pay-when-you-sell publications, either online or in print form, allow you to advertise for free until you sell the item, at which time you pay a percentage of the selling price—sometimes as much as 15 percent. This option is a good idea if you're not sure your item will sell.
- **Visit a reputable dealer in antiques, paintings, guns, jewelry, books, coins, or stamps.** If you think your item has some value, see a dealer who resells the type of item you wish to sell. Don't visit a pawnshop or any other shady business. Go to the best, highest class dealer you can find and present your item for sale. If you aren't satisfied with the price, go elsewhere. That

particular dealer may simply have too many of what you're trying to sell; another dealer may not.

■ **Auction off the item, either online or with a service in your community.** Your items will have to be fairly valuable to others to warrant a live auction (call your local auction company to arrange an appraisal), but even inexpensive items can be auctioned via online services such as eBay (www.eBay.com).

If you choose an option that will require you to send an item to another city or state (and this is usually the case with online auction services), make sure you find out how much FedEx or UPS will charge you to send your item, insured, via one- or two-day service. Add that cost to your base price. Also, do not send your item without first receiving payment: Either send it C.O.D. or require payment from the buyer before sending the item.

Budgeting Tip: *Many online auction services now use a third-party escrow company that receives the money from the buyer and holds it until you send the item and it's received in good condition. The escrow company then sends you the money, charging you, the buyer, or both of you a fee in the process.*

Holding a Tag Sale

If you own a lot of items, but none is of much value, consider holding a tag sale (also called a yard sale or garage sale). Although your items will sell for much less than you paid for them, you may be able to make hundreds of dollars selling items you consider to be junk. Don't forget, however, that lugging all your items out to the garage or yard, marking them with prices, and being anchored to your sale for a day or two is time-consuming and challenging!

Be sure to mark the price on every single item and include a range of prices, from 25 cents for old kitchen towels to $40 for a dresser that's in good condition. This will keep your buyers happy.

To attract customers, set out an attention-grabber—an item that's highly unusual or brightly colored—near the end of your driveway. Make your sale seem full by pulling some of your larger items out of the garage into the driveway. If you don't think you have enough stuff to attract attention, consider combining a sale with neighbors, friends, or family.

Be sure to advertise your sale in your local newspaper. For a fee (generally $10–$25), your sale will be advertised a few days in advance (in the paper and online), and you may even receive some signs to place near your house, at intersections, or on busier streets, showing shoppers how to find you. Your ad should include your address, day(s) of the sale, hours, a list of items, and whether you'll hold your sale in the event of rain. Don't include your phone number in the ad—you'll spend the day of the sale on the phone, distracted from helping your buyers and spotting shoplifters.

Expect early birds to arrive from sixty to ninety minutes before your posted time. If you're not ready to open, ignore them and reiterate that you'll be opening at the time listed in the paper. Most of these early shoppers are antique or resale-shop dealers who want the pick of your tag-sale litter. If you let them in early, regular folks who saw your ad and thought it'd be fun to go to a garage sale may be furious with you!

Good garage sale operators get change (a roll of quarters plus small bills) the day before the sale, but if you do this, keep the money box in your hands at all times. A common scam is for one person to distract you while another steals the money box. Also, if you're not good at addition, keep a calculator nearby.

Cashing In Savings Bonds or Stocks

If you own bonds or stocks that aren't earmarked for your (or your child's) education or for your retirement and they are currently valuable, consider cashing them in to pay down your debt. Before deciding, visit your local bank or stockbroker to determine the value of these assets, as well as any penalties and other costs or commissions associated with selling them.

Using Your House to Pay Off Your Debts

Using your home's equity to get out from under crushing debt was very popular prior to 2008 and the decline in the housing market. Even though it's now done less often, it may be possible to tap your home's equity, which is the value of your home minus the mortgage owed on it.

Declaring Bankruptcy

In 2005, Congress enacted a number of changes to U.S. bankruptcy law. Even though in some cases these changes made it more difficult to file personal bankruptcy, it can still be an option for you if your debt is significant enough and if you have no other options. However, it is not a course to be undertaken lightly.

Bankruptcy generally isn't a good idea because, although it probably seems much easier than credit counseling or selling some of your belongings, it can haunt you for a good portion of your life. Think of it this way: Would you ever loan money to a friend who once borrowed it but never paid it back? Neither will lenders, including those that loan money for cars and homes and those that offer unsecured loans like credit cards and store charge cards. You may even have trouble getting the utilities for your house or apartment hooked up if you've declared bankruptcy. (This means you'll have to prepay these services until you establish a good payment record.) In addition, many stores and other companies won't accept checks from you if you've recently declared bankruptcy.

Requesting Bankruptcy Protection

You can file two types of cases in bankruptcy court, and they're covered in the two following sections.

Chapter 7

In Chapter 7 bankruptcy, nearly all your debts are wiped out; that is, all unsecured debts (credit card balances, hospital bills, long-distance phone bills, and so on) are never paid back. Note that unsecured debts to the government, including student loans, taxes, and court-ordered alimony payments, are not wiped out and must continue to be paid back on an agreed-upon schedule.

Secured debt (cars, mortgage on a house, major appliances) is usually sold, and the proceeds pay off the lender. You may, however, get to keep your house (if you keep paying the mortgage), your car (if you keep paying on the loan, should you have one), and some personal property (TV and so on, as long as you don't owe any money on them). However, you generally will not get to retain an expensive house or car; those will have to be sold.

Many people believe that Chapter 7 bankruptcy is a convenient way to run up a bunch of debts and then walk away, scot-free. Baloney! Chapter 7 bankruptcy is a gut-wrenching heartache that can follow you for at least a decade. Ask yourself this: Why would anyone—especially a creditor who lends money for a living—want to lend you money after you walked away from a pile of unpaid debts in the past? You ate the food, wore the clothes, used the products, and then decided that you didn't want to (or couldn't afford to) pay for them after all. Who would feel compelled to trust you after that? And because you can declare Chapter 7 bankruptcy once every six years, what's going to keep you from doing it again?

Budgeting Tip: Filing for Chapter 7 bankruptcy protection will cost about $300 in court fees. If you hire a bankruptcy attorney, of course, it'll cost you quite a bit more than that—possibly up to $2,000. On the other hand, given the complexity of law on this subject, hiring a lawyer is the safest way to file Chapter 7.

Firms that specialize in bankruptcy insist that new creditors won't know about your past, but that's simply not true. A Chapter 7 bankruptcy can stay on your credit report for ten years. And don't forget that potential employers regularly request credit reports before extending an offer to hire you. They figure it tells them something about the sort of person you are—and they may be right! Even the leasing company at the apartment complex you want to move into and the electric company that's setting up electrical service in your name probably won't agree to work with you if they see a bankruptcy on your credit report.

Chapter 13

Chapter 13 bankruptcy is so much like credit counseling that it should never be your first choice—credit counseling should be. Like credit counseling, you present a plan to the court (including an entire budget that shows that the planned payments are possible) to pay off 100 percent of your debts over as long as five years.

A trustee collects and disperses your payments to creditors, usually charging you an additional 10 percent in the process, an amount that's much higher than what credit counselors charge. You also have to pay about $200 in court fees, and if you use an attorney, you'll have to pay his or her fees. Just about every sort of debt is allowed to be paid off under Chapter 13 bankruptcy, even government loans. Unlike Chapter 7 bankruptcy, with Chapter 13, you usually hold on to your assets.

So why would people choose bankruptcy over credit counseling? Many simply don't know that credit counseling exists, yet chances are, you have a nonprofit credit counseling service right in your city or area. Others believe that bankruptcy is simpler (it isn't) or costs less (it doesn't) than credit counseling. And a few people have had their credit counseling proposals rejected by creditors, and they see bankruptcy as a last option.

According to the Federal Reserve, people filing for bankruptcy typically owe more than one and a half times their annual income in debts (not including their mortgages and cars)! This means that if a family makes $30,000 per year, they owe more than $45,000 in credit card and other high-interest debts.

Think of bankruptcy as your last resort; and if you have to choose, file Chapter 13 protection. But always meet with a credit counseling agency before talking to a bankruptcy lawyer. You'll not only save money, you'll preserve your reputation, too.

11

SAVING FOR RETIREMENT

How much do I need to retire?

This is the million-dollar question: How much money will you need in retirement? Unfortunately, there is no set answer. How much you'll need depends on your expenses during your retirement years as well as what you want to do during those years. This, in turn, is bound up with the long-term goals you set at the beginning of this book.

If you will no longer have a house payment or rent in retirement, you may be able to live on a lot less than if you continue to make those payments. On the other hand, if you live in an older house in retirement, you may also encounter more repairs than someone in a new condo. In the same way, you may want to belong to a country club in retirement, but get rid of your expensive car. So, how much you need depends on your individual circumstances.

Use Worksheet 11-1 as a way to begin to determine what your expenses will be during your retirement years. Think of how you live your life now, what will likely change in retirement, and what will be paid off before you retire. (Hint: Don't worry so much about what the actual costs might be down the road; just write in what those costs would be in today's dollars to give you a handle on how your expenses will increase or decrease in retirement.)

Worksheet 11-1: Retirement Expenses

Monthly Expense	Amount
Groceries and household items	$
Contributions	$
Savings	$
Rent on furniture or appliances	$
Entertainment	$
Eating out	$
Rent or mortgage	$
Car payment or lease	$
Electric bill (average)	$
Gas bill (average)	$
Sewer bill	$
Water bill	$
Trash pickup bill	$
Cable/DSL/satellite bill	$
Telephone bill	$
Cell phone bill	$
Bank charges	$
Haircuts/manicures/pedicures	$
Home equity loan	$
Other loan	$
Credit card or store-charge card bill	$
Credit card or store-charge	$

card bill	
Credit card or store-charge card bill	$
Credit card or store-charge card bill	$
Credit card or store-charge card bill	$
Credit card or store-charge card bill	$
Child support or alimony	$
Car maintenance	$
House maintenance	$
Auto insurance	$
Property taxes	$
Gifts	$
Events to attend	$
Clothing and shoes	$
Home insurance	$
Vehicle registration	$
Vacation	$
Club membership	$
Club membership	$
Club membership	$
TOTAL:	$
Other:	$
Other:	$

Finding Ways to Set Money Aside Now

The main reason people put off saving for retirement is that they think they have plenty of time to do that later. The second most popular reason for putting it off is that most people don't know how to come up with money to put into savings. If you've read any of the chapters in the first third of this book, though, you can probably come up with a variety of ways to find $50, $100, or $200 a month for retirement savings. Use Worksheet 11-2 to brainstorm ideas, and review the following sections for some creative ways to save more for retirement.

Worksheet 11-2: Ideas for Reducing Your Current Expenses

Expense	Idea for Reducing or Eliminating	Potential Monthly Savings
		$
		$
		$
		$
		$
		$
		$

Here are some ideas to save money, money that you can put aside for retirement.

Stop Eating Out

If you eat takeout twice a week, and you pay $12 for a meal that you could make for $3.50 at home, you could put about $74 a month into your retirement savings. Over twenty years at 5 percent, that's $18,564.

Cut Your Clothing and Shoe Budget in Half

If you spend $1,000 per year on clothing and shoes, can you cut that amount in half and put $500 a year ($42 per month) into a retirement account? Thirty years of that, at 6 percent, and you'll have $42,189.63 toward your retirement fund.

Move to a Smaller House

If you're currently living with two other people in 2,400 square feet, could you move to a house with 1,800 square feet and still be comfortable? If so, your mortgage payments might go down by anywhere from $300 to $1,000 per month—money that you could put into a retirement fund. In just fifteen years at 6 percent, a $600-per-month savings will equal a whopping $157,382.85.

Drive Your Car Twice as Long

If you currently get a new car every three years, pay it off, and get another new one, try something different: Pay off your car in three years but drive it for six. Put the amount of your car payment into a retirement account for the second three years. You'll only contribute to your retirement account three years out of every six, but you'll have found a creative way to save.

Start a Part-Time Business

Instead of looking only at potential expenses to cut, consider working a few extra hours per week, perhaps at your own business, and putting that income toward your retirement savings.

Looking at Tax-Deferred Ways to Save

When most people think of retirement and the government at the same time, they think of Social Security, the government program that collects money from you throughout your working life and gives it back to you, one month at a time, during your retirement years.

If you're nearing retirement, you can probably count on quite a bit of Social Security income for your retirement. If you haven't already, you will soon receive a statement that explains how much you'll receive each month, based on exactly what age you retire. You can also contact the Social Security office at www.ssa.gov or (800) 772-1213.

If you're forty or younger, however, there's isn't much chance that Social Security will fully fund your retirement. That's because instead of taking income from you, investing that income, charging a small administrative fee, and then paying you benefits from the investment, as you may have expected the federal government to do, the system actually works quite differently. The money you paid in isn't there anymore: What you paid in ten years ago was used to pay benefits to other people ten years ago; what you paid in last week was used to pay benefits to other people last week. That worked pretty well when the largest generation in American history (the baby boomers) was working, but as boomers enter retirement, the government will probably not be able to collect enough in Social Security income to offset the benefits being paid to them.

As a result, the government has taken two steps: increasing the age at which you can begin to draw Social Security benefits; and encouraging you to invest more money in your own retirement accounts to use when you retire. The following sections give you some examples of how they hope to encourage you to do that.

As you investigate all of your options, use Worksheet 11-3 to keep track of what's available to you and how much you're legally allowed to contribute.

Worksheet 11-3: Possible Retirement Savings

Type of Account	Possible Contribution	Employer Matching Amount
		$
		$
		$
		$
		$
		$
		$
		$

Starting Young

The absolute best way to save for retirement is to start young. If you start saving $100 per month when you're twenty-five, and you invest that in a mutual fund or other stock-related fund that sometimes earns 18 percent interest and sometimes loses money, averaging 8 percent over the next forty years, you'll have $351,428.13 for retirement. To get the same amount of retirement savings if you start at age forty, you'll have to put away about $367 per month. Use the Savings Growth Projector calculator at www.finaid.org to run these numbers for yourself.

Making Up for Lost Time

Regardless of how old you are, you can still save some money for your retirement, thanks to the government's catch-up provision, which allows you to save more money in a retirement account when you're fifty or older than when you were younger. You may also be able to use the equity you've built up in your house to fund part of your retirement. The following two sections share some details about each option.

Using the Catch-Up Provision

If you're fifty or older, you're allowed to put away more tax-free income per year than your younger counterparts. For example, while the standard contribution to both traditional and Roth IRAs is $5,000 per year (subject to income limits, of course), anyone fifty or older can contribute $1,000 more than that. Your company's 401(k) plan has similar provisions, allowing you to contribute more than younger workers. This is called a catch-up provision, and it's meant specifically for people who started saving for retirement later in life.

Using Your House to Help You Retire

If you haven't made many provisions for your retirement and can't seem to find the money to do so, you may be able to tap the value of your house for your retirement. Suppose you're forty-five years old and have fifteen years left to pay on your house. You bought the house fifteen years ago, and home prices have risen significantly since then. Your house is worth $240,000 now and will likely be worth over $450,000 when you turn sixty and the house is paid off.

Rather than staying in the house during retirement, you can sell it and move to a smaller house or condominium that costs far less. If you can sell your house for $450,000 and buy a condo at that time for $300,000 or $350,000, you'll have $100,000 to $150,000 to add to your retirement account.

Budgeting Tip: *If mortgage interest rates drop to one point or lower than the rate on your current mortgage, consider refinancing. You may be able to get a shorter loan length (say, fifteen years instead of thirty) for the same monthly payments. This can enhance your ability to use your house as part of your retirement plan.*

On the other hand, you can stay in the house and get a reverse mortgage on it as soon as you begin needing income (say, when you turn sixty-five or seventy). Essentially, a bank buys the house back from you, except that you continue to own it and live in it. You must be at least sixty-two and own the house free and clear. The bank then either sends you monthly payments or gives you a lump sum. You'll pay a fee for this service because if you live longer than the bank thinks you're going to live, they may actually lose money on the deal. Still, many lenders offer this option.

CONCLUSION

Too many people, when planning a budget, decide that to meet their financial challenges they must cut absolutely everything. They stop eating and drinking as much, they eliminate visits to the hairdresser, they cut out shopping for clothing, and they never, ever go to the movies.

Such well-intentioned budgeting doesn't usually last very long. First one thing slips, then another, and before you know it the budget is just a piece of paper, shoved to the back of a drawer and forgotten.

The key to successful budgeting, as I've stressed in this book, is that you've got to want to do it. You've got to be motivated, disciplined, and sensible. This involves the following basic steps:

- <u>Decide what your dreams are.</u> These can be short-term or long-term; it doesn't matter. What's important is that they matter to you. The more you care about your goals, the easier you'll find it to stick to a budget.

- <u>Figure out where you're at</u>. In Chapters 3, you looked realistically at what your income and outgo are and what you owe in short- and long-term debt.
- <u>Decide where you can make savings</u>. The savings you make should be realistic, motivated by your goals, and easily achievable.

- Take Action. Decide ONE THING you can absolutely cut from your life. Whatever you decide, go at least 60 days without it. Run this experiment every single month.
- Invest in your budget. If the budget is just for you, review it regularly and adjust it as needed. Keep your goals front and center; they're what drive you to stick to the budget. If you're part of a family, let the other family members in on the budget. Get ideas from everyone about savings and how to meet your dream goals.

Don't be surprised if your goals shift quite a bit over time. For example, retirement today is much more expensive proposition than it was for many of our parents. That's natural; life's gotten more expensive.

Budgeting isn't a one-time event and a budget isn't a rigid, fixed document—it's a living thing. It grows and changes along with you. Don't be afraid to constantly review it and make sure it's working and that it's getting you closer to your goals. Above all, view it as a tool— one that will get you from where you are now to where you want to be.

> *"You only HAVE TO do something until you WANT TO...then you DON'T HAVE TO anymore"*
> -Leland Val Van De Wall

You will always want to work on increasing your income and being frugal at the same time. Think positively and believe in yourself. That way, you will get one step closer to your financial freedom and be on the fast track towards financial independence!

THANK YOU

Thank you for purchasing and reading this book all the way to the end. I really hope this book was able to help you.

If you enjoyed this book...

I would like to ask you a small favor: Could you take a moment and please share your thoughts in a REVIEW?

They really do make a difference and are a great help to other readers to make a decision. Positive reviews are very valuable and helpful and I would love to hear from you! They really mean a lot to me and I would really appreciate it.

Thank you so much!
Rachel Mercer

Here is the URL address of the book page on Amazon to leave your review.

www.amazon.com/dp/B086DK2DXV

YOUR FREE GIFT

If you haven't done so already, please accept my **FREE special PDF report, "11 Best Side Hustles You Can Do Anywhere at Any Time to Make Extra Money"**. It will give you a simplest and easiest side hustle ideas to increase your income and tips on how to grow your wealth. You'll be surprised as to what you'll find out in this guide – as side hustles can definitely level up your finances.

These little things every month will go a long way. In several months or in a few years' time, you'll be surprised as to how much you had actually saved through little increments that accumulated into something bigger.

Get your free gift here:

https://bit.ly/39hoopf

Free downloable worksheets can be found here:

Http://BookHip.com/GPFXRX

Printed in Great Britain
by Amazon